McGraw-Hill

Mathematics

McGraw-Hill
School Division
New York Farmington

Senior Consulting Authors

Gunnar Carlsson, Ph.D.
Professor of Mathematics
Stanford University
Stanford, California

Ralph L. Cohen, Ph.D.
Professor of Mathematics
Stanford University
Stanford, California

Program Authors

Douglas H. Clements, Ph.D.
Professor of Mathematics Education
State University of New York at Buffalo
Buffalo, New York

Carol E. Malloy, Ph.D.
Assistant Professor of Mathematics Education
University of North Carolina at Chapel Hill
Chapel Hill, North Carolina

Lois Gordon Moseley, M.S.
Mathematics Consultant
Houston, Texas

Robyn R. Silbey, M.S.
Montgomery County Public Schools
Rockville, Maryland

McGraw-Hill School Division
A Division of The **McGraw-Hill** *Companies*

McGraw-Hill School Division
Two Penn Plaza
New York, New York 10121-2298

Printed in the United States of America
ISBN 0-02-100123-5
9 073/043 05

Contributing Authors

Mary Behr Altieri, M.S.
Mathematics Teacher
1993 Presidential Awardee
Lakeland Central School District
Shrub Oak, New York

Nadine Bezuk, Ph.D.
Professor of Mathematics Education
San Diego State University
San Diego, California

Pam B. Cole, Ph.D.
Associate Professor of
Middle Grades English Education
Kennesaw State University
Kennesaw, Georgia

Barbara W. Ferguson, Ph.D.
Assistant Professor of Mathematics
and Mathematics Education
Kennesaw State University
Kennesaw, Georgia

Carol P. Harrell, Ph.D.
Professor of English and English Education
Kennesaw State University
Kennesaw, Georgia

Donna Harrell Lubcker, M.S.
Assistant Professor of Education
and Early Childhood
East Texas Baptist University
Marshall, Texas

Chung-Hsing OuYang, Ph.D.
Assistant Professor of Mathematics
California State University, Hayward
Hayward, California

Contents

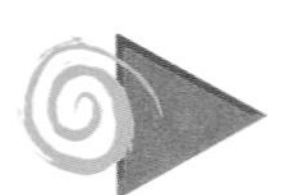

Readiness

Theme: Ready, Set, Go!

Chapter 1: Position and Classify

Theme: Getting to Know You

Chapter 2: Sorting

Theme: I Spy!

Chapter 3: Data and Graphs

Theme: Show and Tell

Chapter 4: Patterns

Theme: Going on a Safari

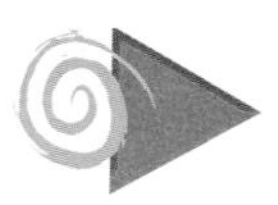

Chapter 5: Numbers to 5

Theme: Out on the Farm

Chapter 6: Numbers to 10

Theme: Hops, Skips, and Jumps

Chapter 7: Numbers to 20

Theme: Around the Neighborhood

Chapter 8: Numbers to 100

Theme: Fantastic Adventure

Chapter 9: Money

Theme: Going Shopping

Chapter 10: Measurement

Theme: Construction Site

Chapter 11: Time

Theme: Going Places

Chapter 12: Addition Concepts

Theme: In the Garden

Chapter 13: Subtraction Concepts

Theme: Flying Friends

Chapter 14: Geometry and Fractions

Theme: Snack Time

Readiness

START

1 2 3

theme
Ready, Set, Go

Use the Data

What do you think the turtles are doing?

What You Will Learn

In this chapter you will learn how to:

- Identify when something happens.
- Identify how often something happens.
- Name the days of the week.
- Identify circle, square and triangle.

Dear Family,

As the school year begins, I will learn my phone number and address. I will also learn the days of the week as well as the names of shapes. Here are some new vocabulary words and an activity that we can do together.

Shape Search

- Open a magazine or book to a page with a variety of pictures.

- Find a square, circle or triangle on the page and give clues to help your child find it. For example, say "I spy a red circle on a street sign."

use

magazine

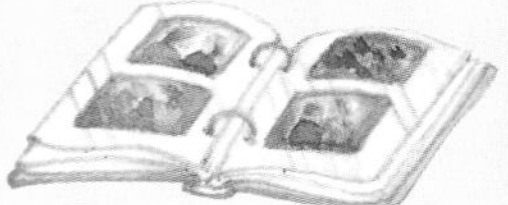

photo album

picture book

Math Words

circle

square

triangle

Additional activities at www.mhschool.com/math

Name ______________________________

Math Words
all
some
none

all some none

Use cubes. Place cubes in the jars to show all, some and none. Draw the cubes.

Math at Home: Your child used cubes and colored objects to show all, some, and none.
Activity: Fill a small bowl with macaroni. Put all, some or none of the macaroni on a plate. Ask your child whether the plate has all, some, or none of the macaroni.

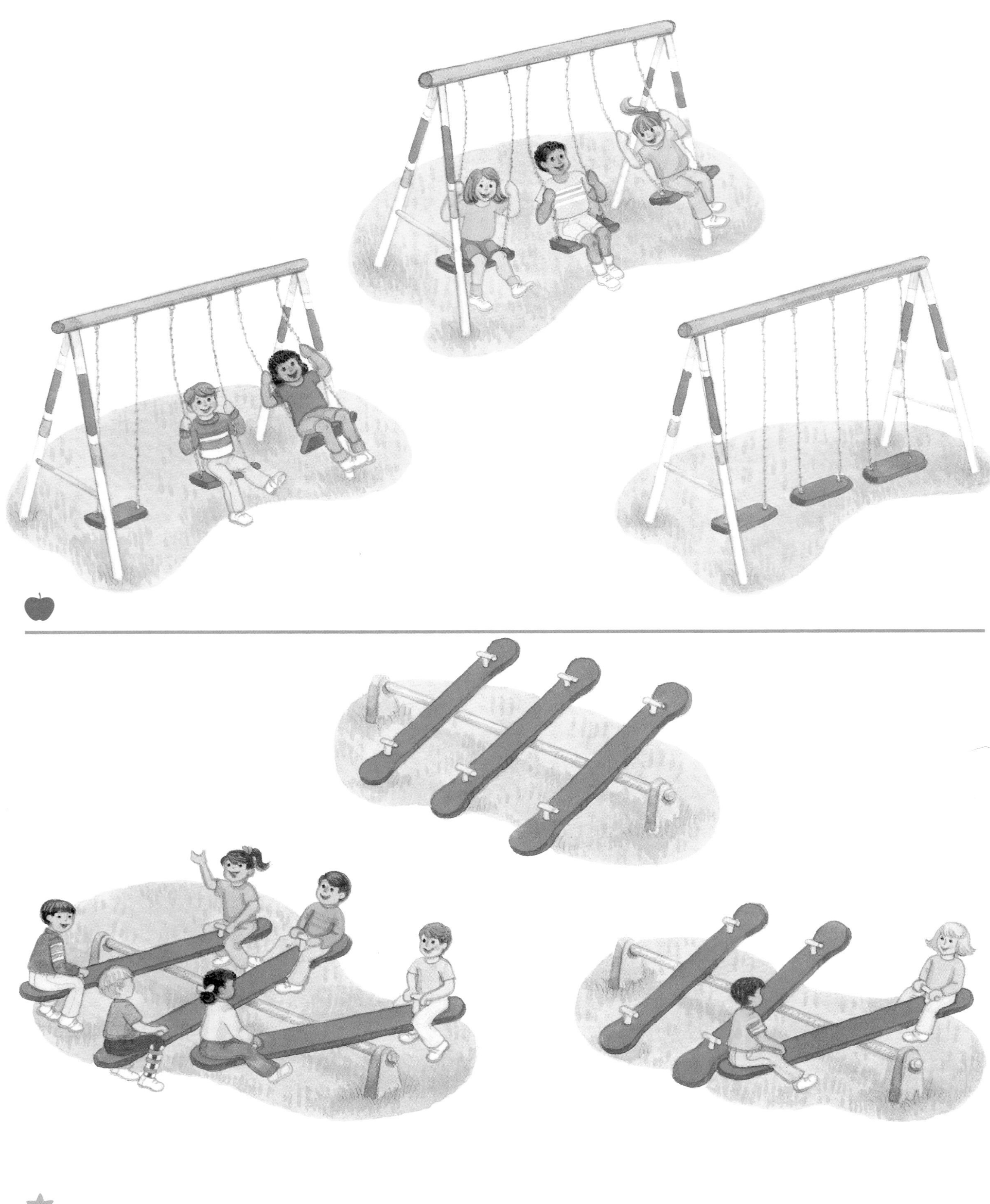

Circle each picture that shows all of the playground equipment being used.
Draw a line under each picture that shows some of the playground equipment being used.
Draw an X through each picture that shows none of the playground equipment being used.

Name____________________

Math Words
always
maybe
never

always

maybe

never

Color a cube to show what you will always, maybe and never pick.

Math at Home: Your child used the words always, maybe and never to tell if something would happen.
Activity: Ask your child if certain events will always, maybe or never happen. For example, the sun rising, or seeing a rainbow.

Draw a red circle around things that will *always* happen.
Draw a blue circle around things that will *never* happen.
Draw a green circle around things that will *maybe* happen.

Name ____________________

Before and After

Draw a picture to show something that happens before or after you wake up in the morning.

Math at Home: Your child drew pictures to show before and after.
Activity: Ask your child before and after questions about events at home. For example, ask what meals are before and after lunch.

Look at the first picture in each row.
Draw a red circle around something that happened before.
Draw a blue circle around something that happened after.

Name ____________________

Say these words.

Math Words
day
night

Day

Draw pictures to show things that happen during the day and things that happen at night.

Math at Home: Your child drew pictures to show day and night.
Activity: Have your child tell you about one thing he or she does in the day and night.

Which events usually happen during the day? Draw a blue circle around them.
Which events usually happen at night? Draw a red circle around them.

Name ____________________

today

Math Words

yesterday

today

tomorrow

Say these words.

yesterday

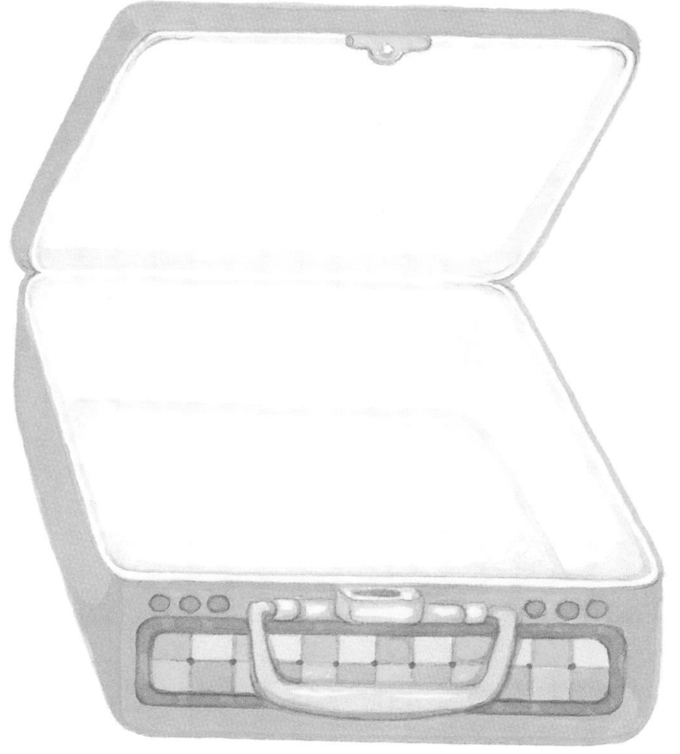

tomorrow

Draw a picture of something you will eat for lunch today. Draw a picture of something you ate for lunch yesterday. Draw a picture of something you want to eat for lunch tomorrow.

Math at Home: Your child drew pictures about yesterday, today and tomorrow.
Activity: Ask your child what day of the week today is. Then have your child name yesterday and tomorrow.

tomorrow

This calendar shows the fruit Jack eats for lunch each day. Draw a red circle around today. What did Jack eat today? ★ What day was yesterday? Draw a picture to show what Jack ate yesterday.

What day is tomorrow? Draw a picture to show what Jack will eat tomorrow.

Name______________________________

Algebra & functions

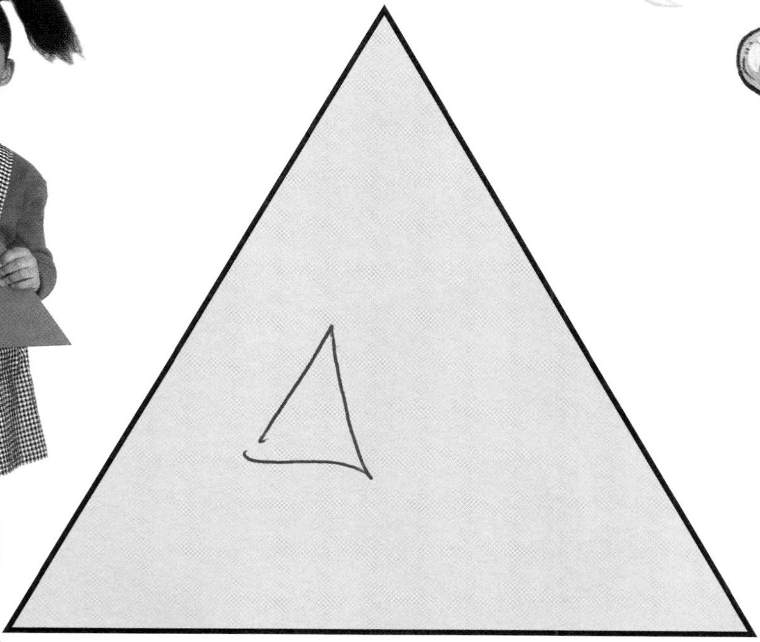

Say these words.

Math Words
circle
square
triangle

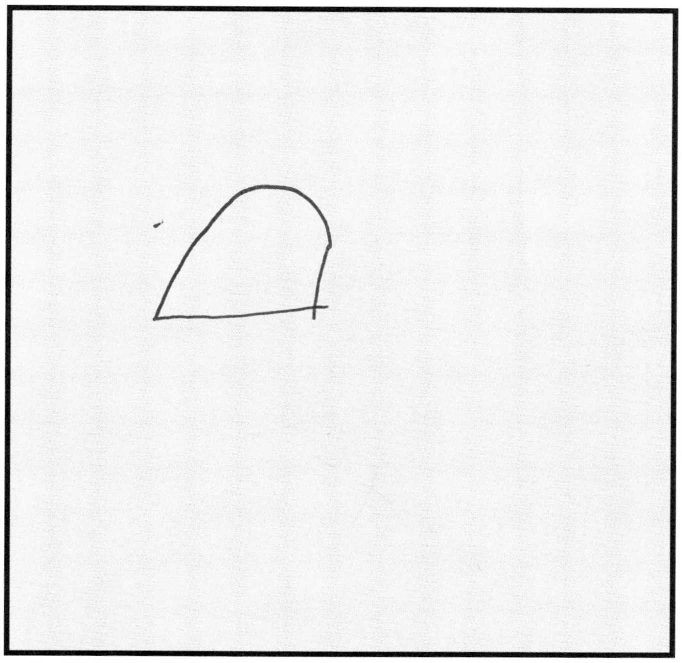

triangle

square

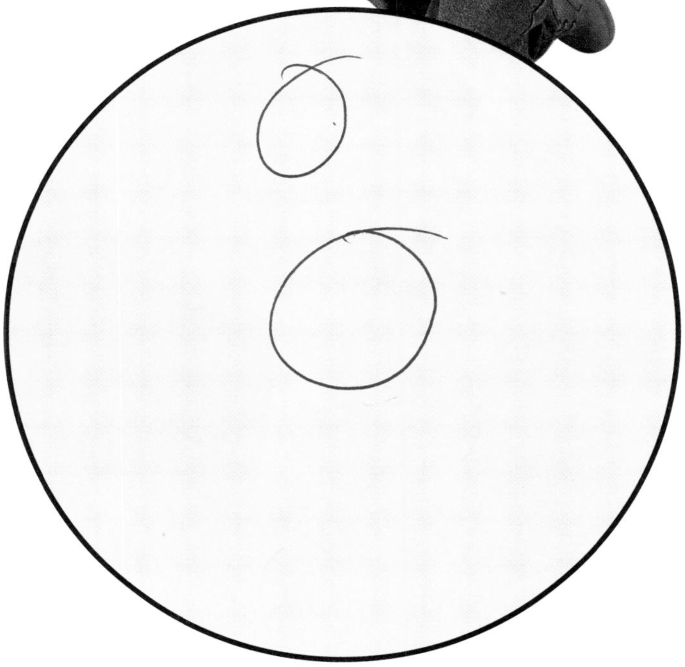

circle

Use attribute blocks to sort circles, squares and triangles.
Draw the attribute blocks to show the shapes.

Math at Home: Your child worked with shapes and drew pictures to show circles, squares and triangles.
Activity: Ask your child to draw you a picture using circles, squares and triangles.

Find the shapes. Color.

CHAPTER 1

Position and Classify

theme
Getting to Know You

Use the Data

What do you notice about the children?

What You Will Learn

In this chapter you will learn how to:

- Use position words.
- Tell how objects are the same or different.
- Use clues to solve problems.

Dear Family,

In Chapter 1, I will learn about position words and tell how objects are the same or different. Here are some new vocabulary words and an activity that we can do together.

Where is It?

Math Words

top

middle

bottom

inside outside

same different

- Choose an item and ask your child to identify it. Use *top, middle* or *bottom* as you ask your child to put the item on a shelf. "Put the ball on the bottom shelf."

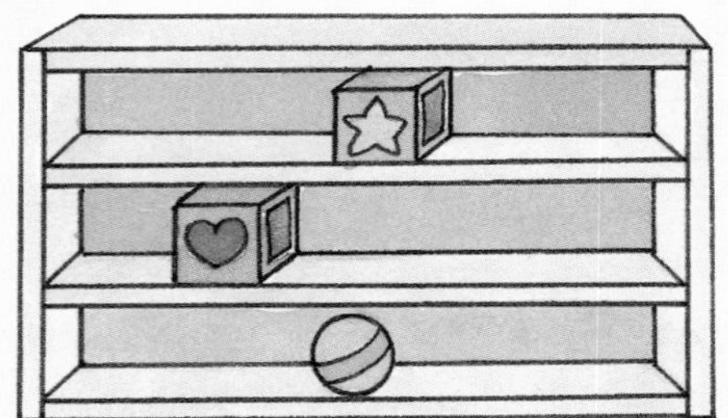

- Continue with the other items, varying the positions.
- Then name the items one at a time. Ask, "Which shelf?"

use

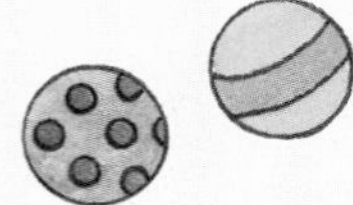

balls

or stuffed animals

or blocks

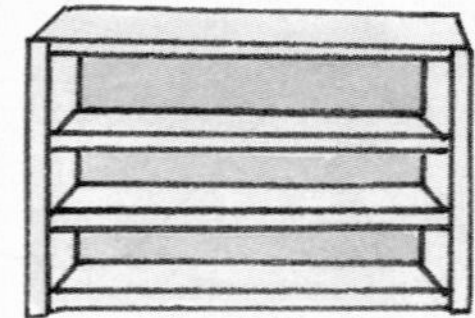

and a bookcase

or boxes

Additional activities at www.mhschool.com/math

Name ______________________________

Say these words.

Math Words
top
middle
bottom

Listen to your teacher. Place cubes on the top, middle, and bottom shelves. Color the object on each shelf to match the cube.

Math at Home: Your child used cubes and colored objects to show the positions top, middle, and bottom.
Activity: Look at pantry or refrigerator shelves with your child. Ask your child to name items that are on the top, middle, and bottom shelves.

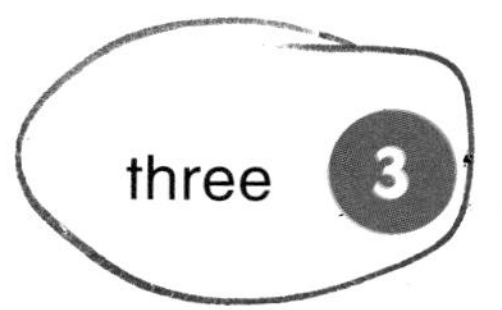

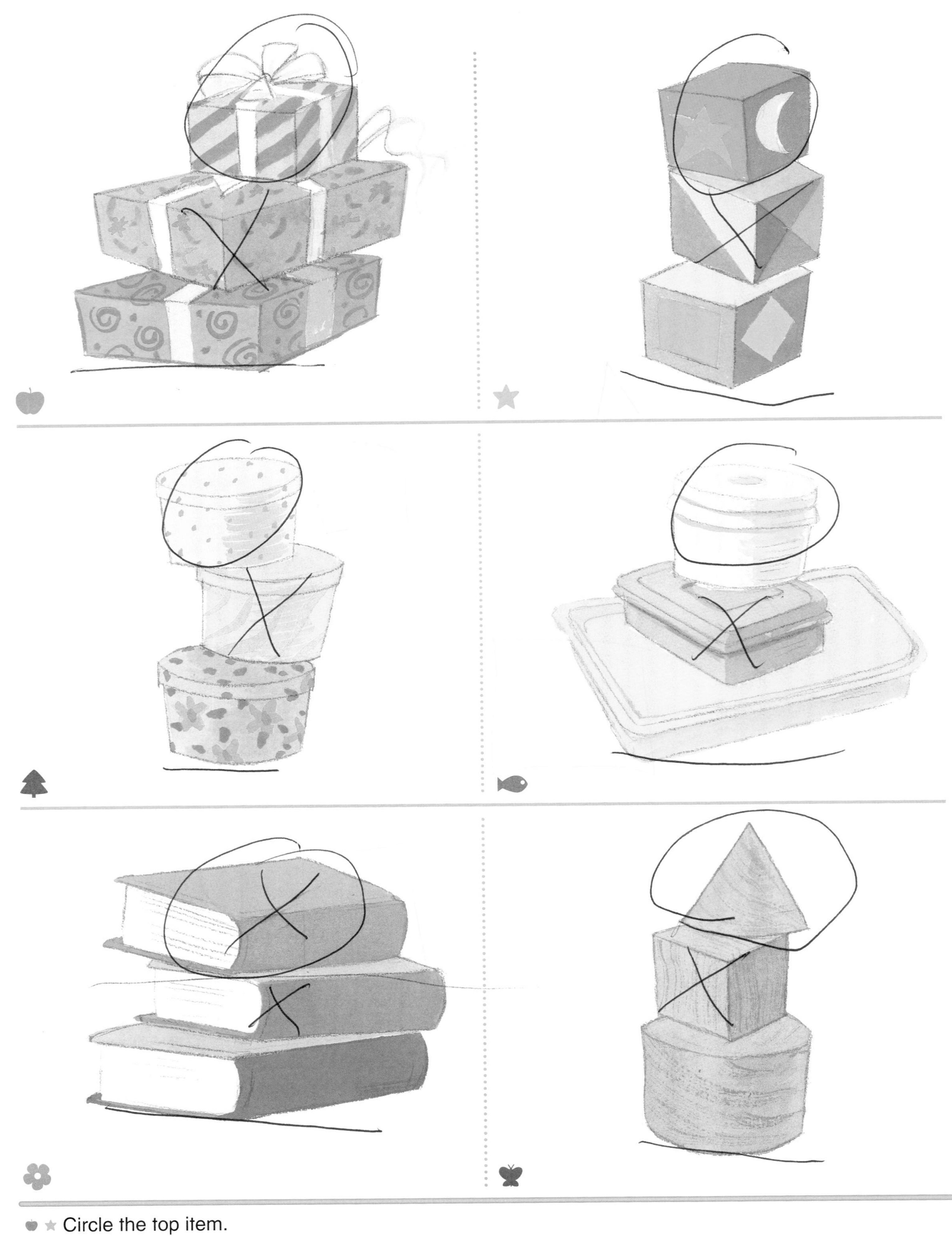

Circle the top item.

Draw an X on the middle item.

Draw a line under the bottom item.

Name Ciara

Math Words
inside
outside

Say these words.

Drop cubes on the page. Tell which cubes are inside the box. Tell which cubes are outside the box.
Draw a pail inside the box. Draw a shovel outside the box.

Math at Home: Your child used cubes and drew pictures to show the positions inside and outside.
Activity: Ask your child to name items inside his or her room. Then ask him or her to name items outside his or her room.

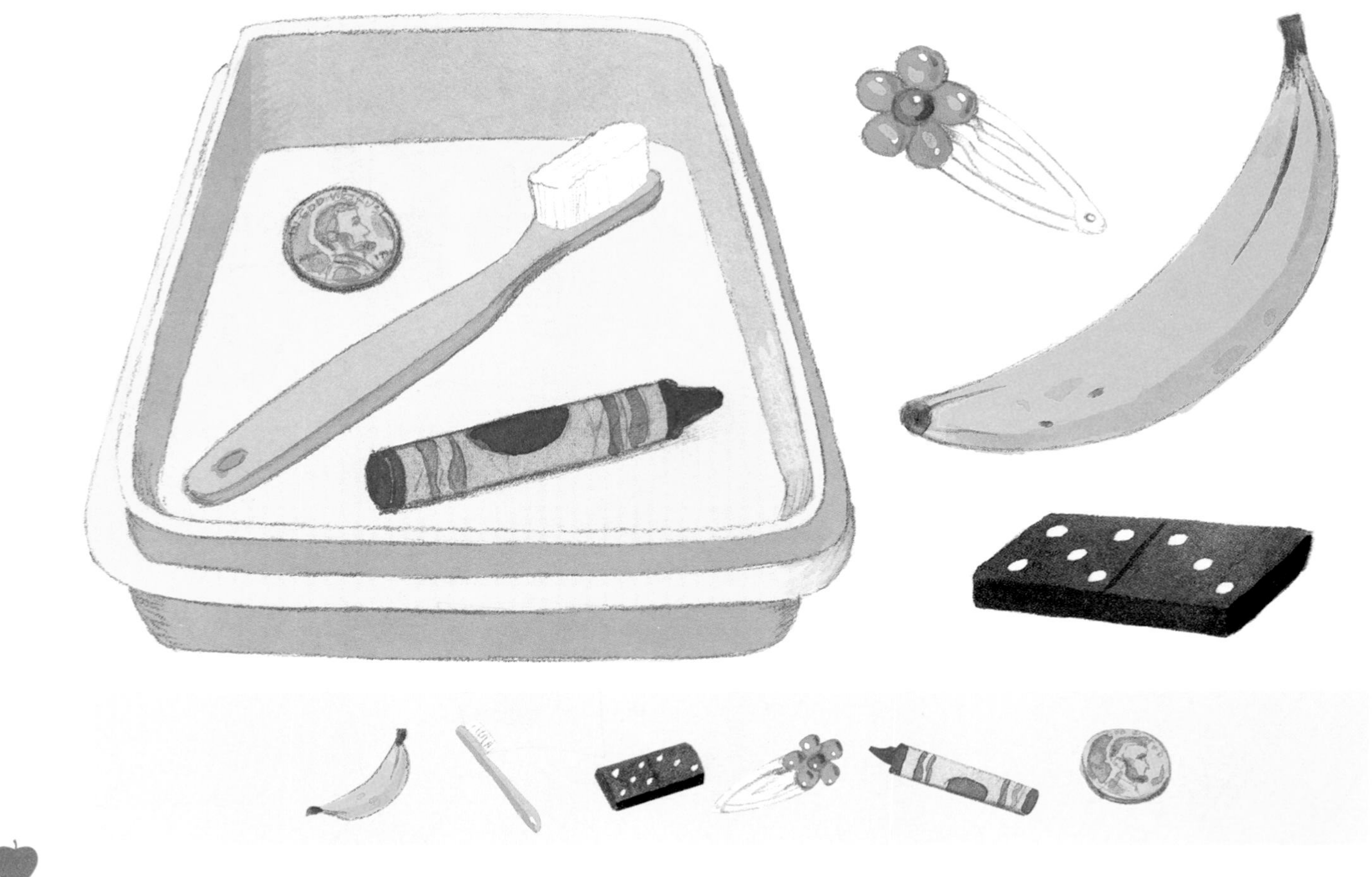

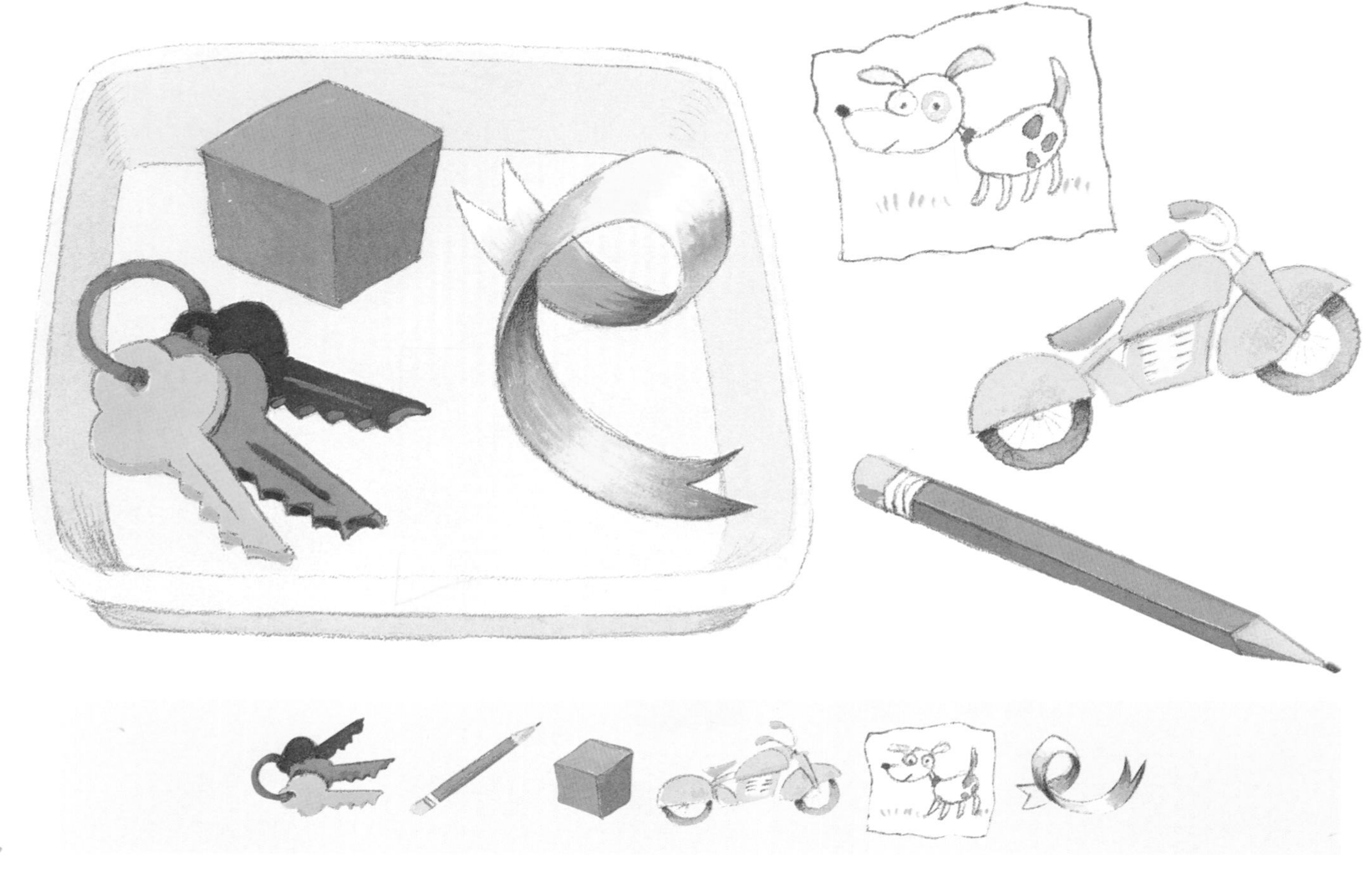

- Circle the objects you see inside the box.
- Circle the objects that you see outside the box.

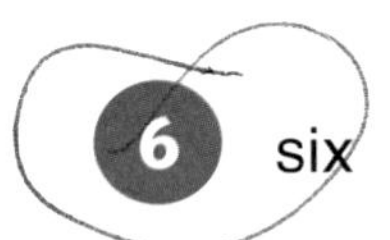

Name ____________________

Say these words.

Math Words

over
under
behind

Use cubes to show over, under, and behind.
Draw a shovel behind the pail. Draw a bird over the swings. Draw a feather under the slide.

Math at Home: Your child drew pictures to show the locations over, under, and behind.
Activity: Ask your child questions about the locations of objects in a room. For example, ask "What is behind the salt shaker?", "What is over the table?" or "What is under the sink?"

Draw a red circle around bunnies that are behind something.
Draw a blue circle around bunnies that are over something.
Draw a yellow circle around bunnies that are under something.

Name______________________________________

left

right

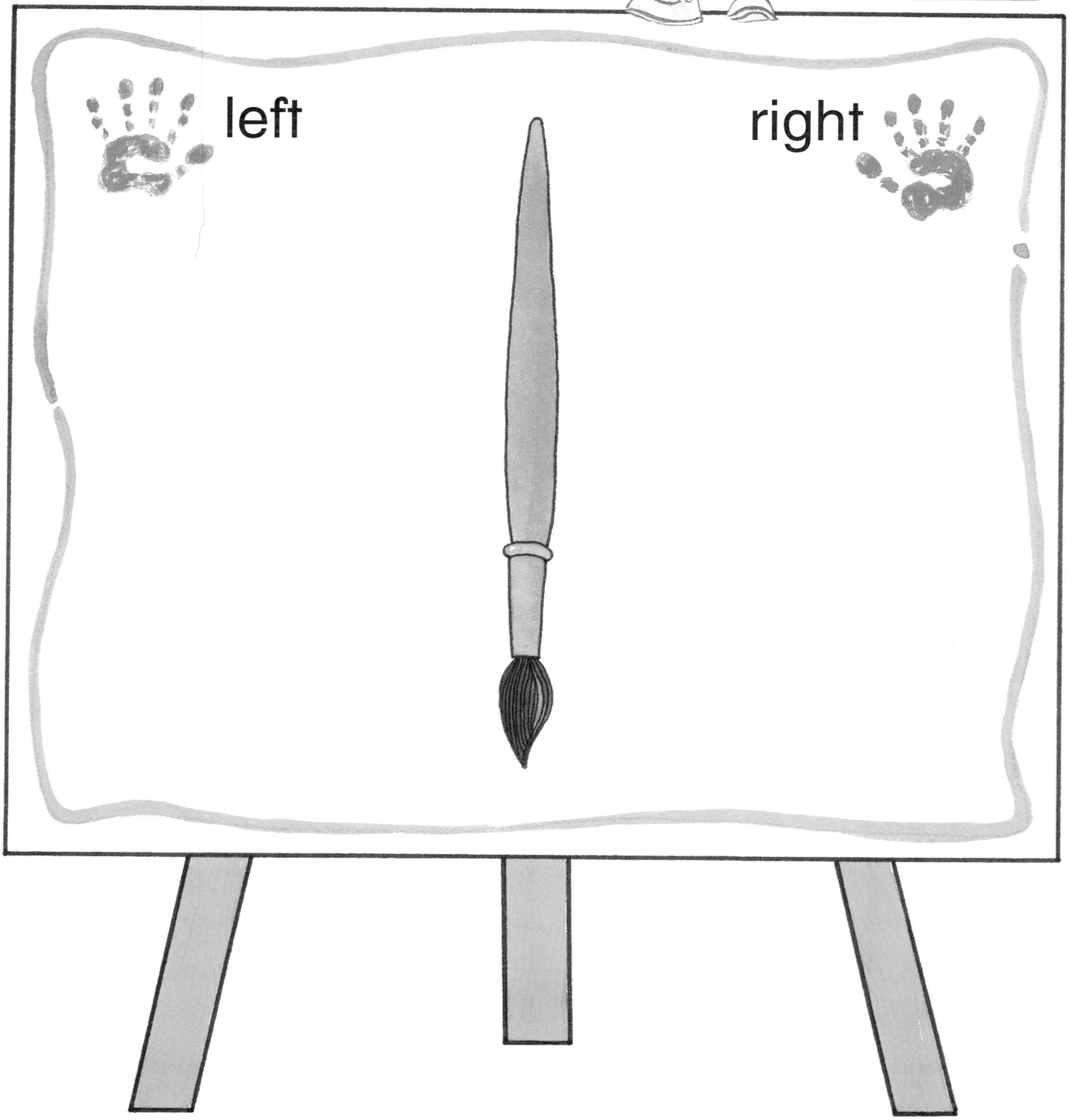

Listen to your teacher. Place cubes on the left or right side. Draw a picture of something you wear on the left side of the mat. Draw a picture of something you eat on the right side of the mat.

Math at Home: Your child used cubes and drew pictures to show the positions left and right. **Activity:** Place a plate, butter knife, fork, and spoon on the table. Ask your child to put the knife on the right side of the plate, the spoon to the right of the knife, and the fork on the left side of the plate. Then have your child describe the location of each utensil.

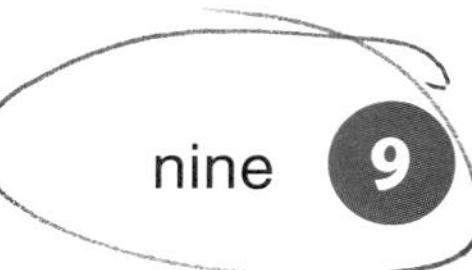

- Circle the items you see on the easel on the left.
- Circle the items you see on the easel on the right.

Name___________________________

Algebra
& functions

Use Logical Reasoning

Listen to the clues. Find the picture. Color.

Math at Home: Your child used clues to solve problems.
Activity: Draw a series of faces on a sheet of paper. The faces should be the same except for one or two different features (smiling, frowning, one eye closed, curly hair, etc.). Give clues about each face, and have your child find the face.

Listen to the clues. Circle the face.

Name ______________________________

Algebra & functions

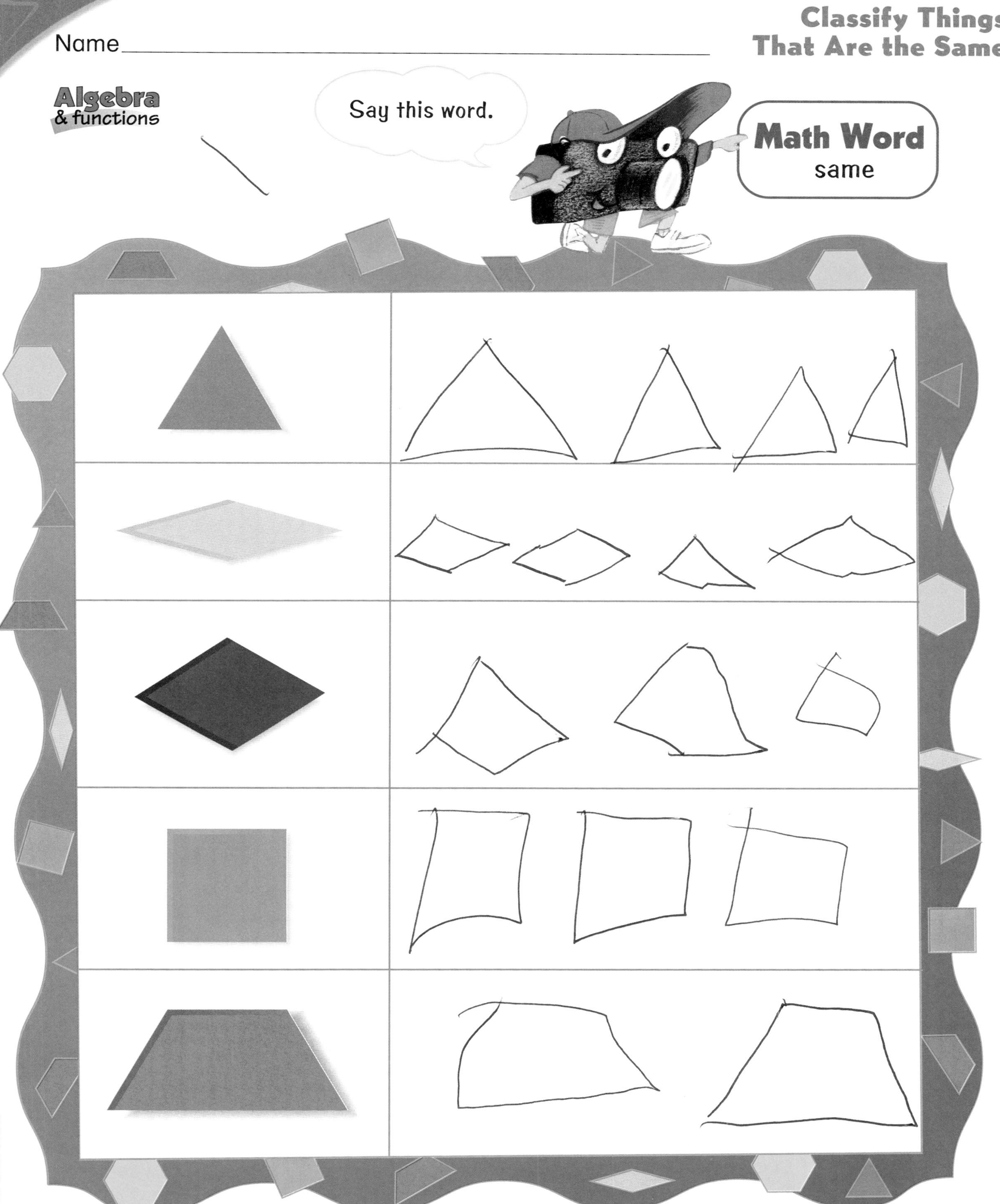

Find each pattern block. Place the pattern block next to the picture. Trace each pattern block.

Math at Home: Your child found objects that were the same size, color, or shape.
Activity: Help your child collect an assortment of leaves or buttons. Then ask your child to sort them in various ways, such as by size, color, or shape.

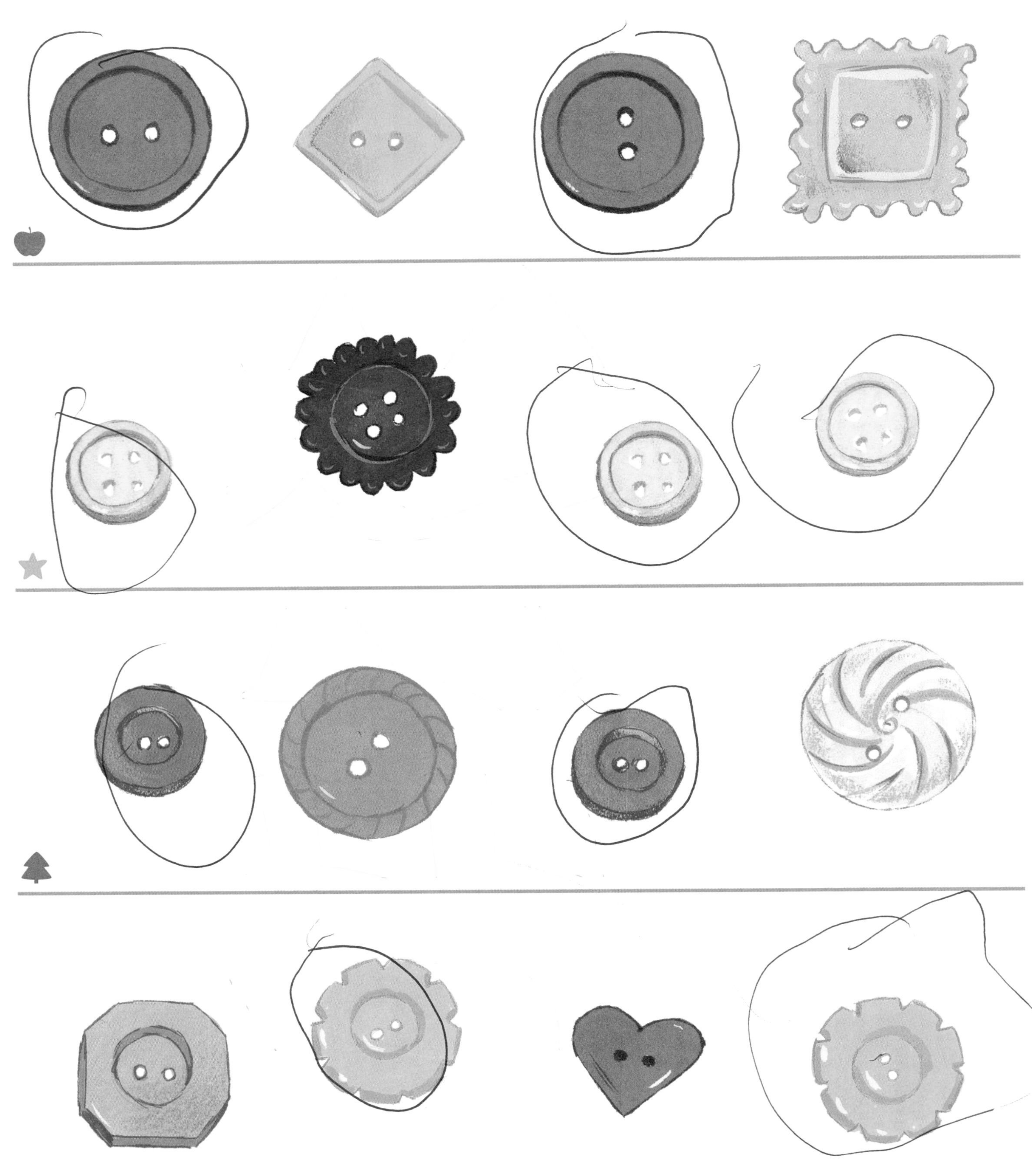

Circle the buttons that are the same.

Name ________________________________

Algebra & functions

Say this word.

Math Word
different

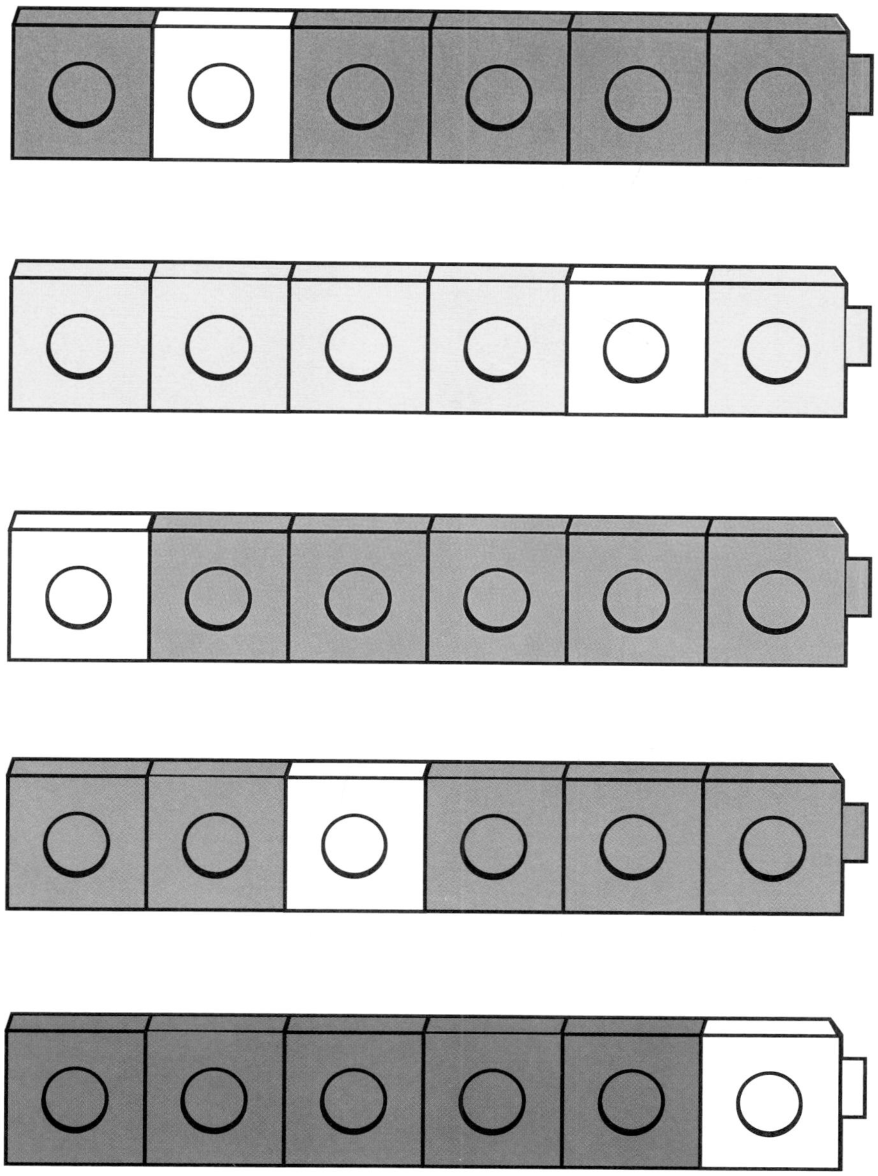

Color a cube in each row to show something that is different.

Math at Home: Your child colored cubes to show objects that do not belong to a group.
Activity: Make a group of three objects that are the same and one that is different, such as three spoons and a fork. Ask your child to find the object that does not belong.

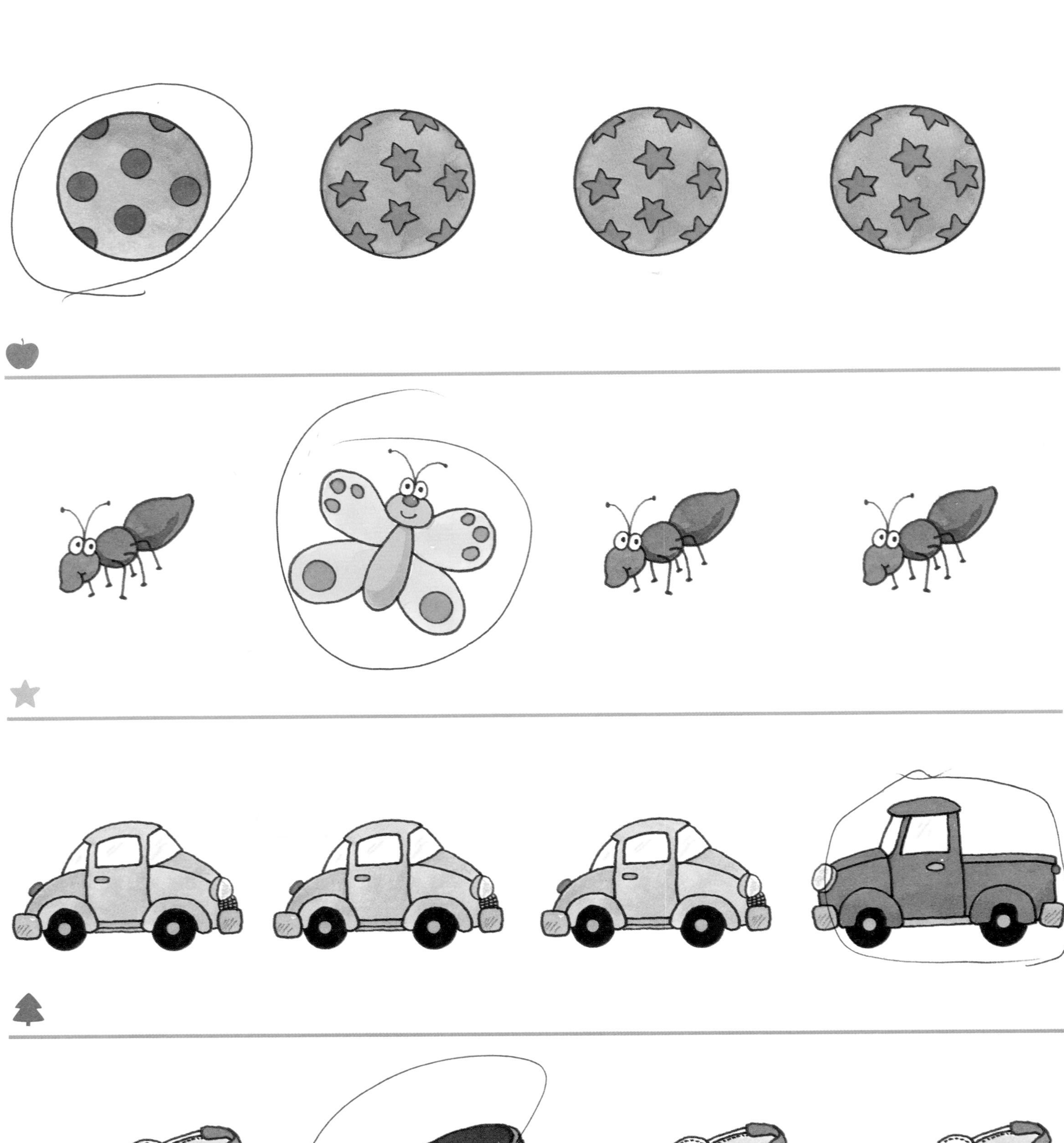

Draw an X on the object in each row that is different.

Name ______________________________

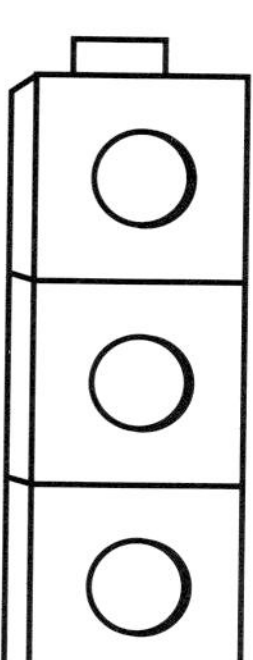

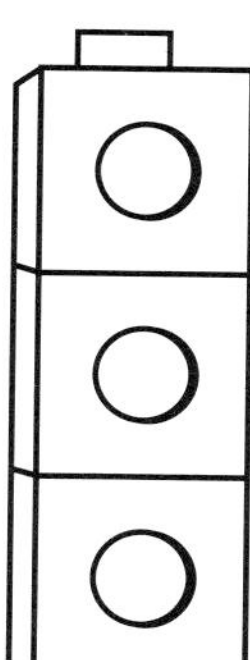

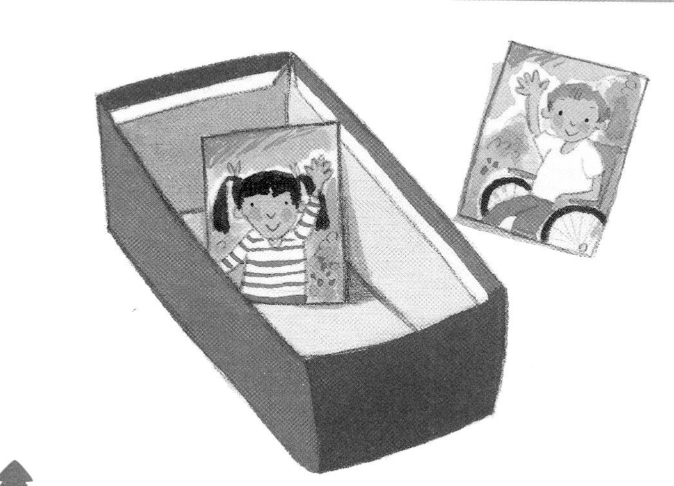

Color the middle cube yellow. Color the top cube blue.
Draw a circle around the picture inside the box. Draw an x on the toy outside the box.
Color the fish over the seaweed blue. Color the fish behind the house green.

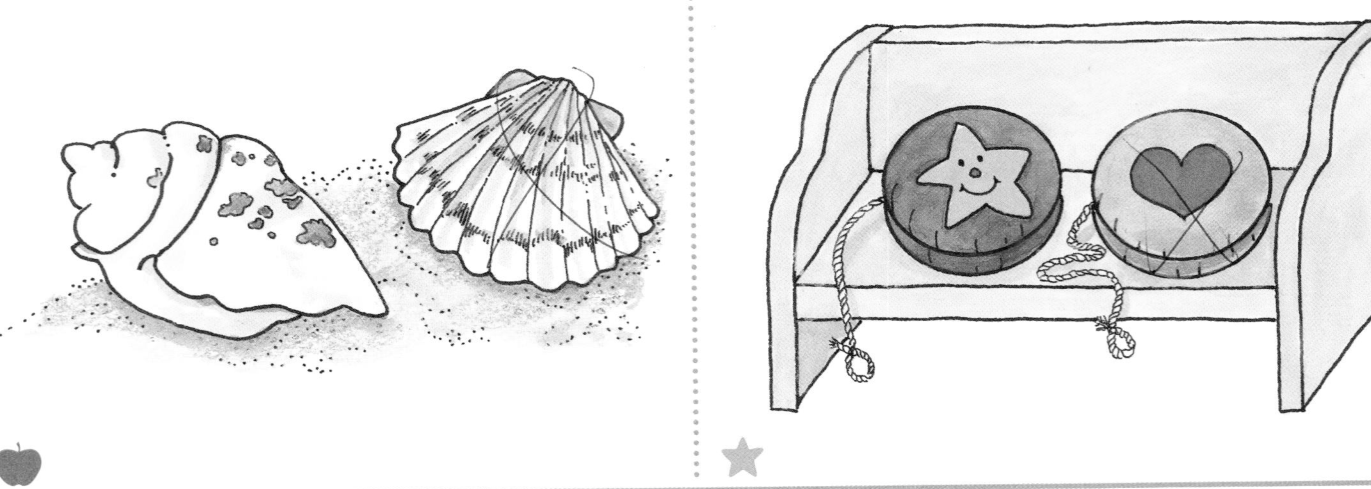

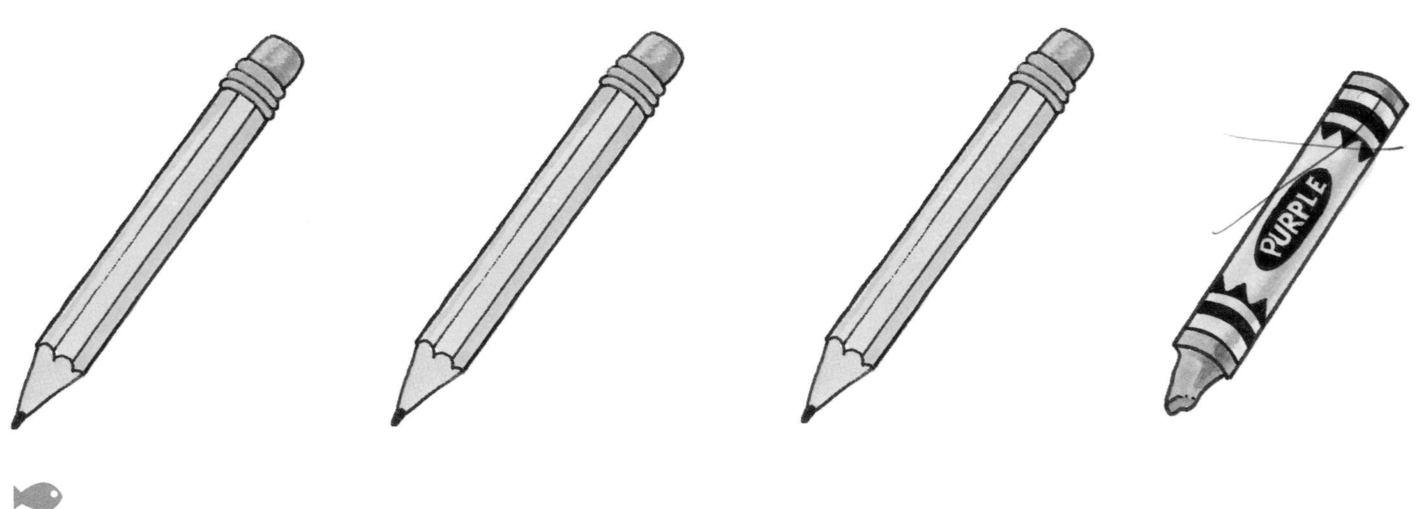

Draw an X on the seashell that is on the right. Draw a circle around the yo-yo that is on the left.
Circle the items that are the same. Circle the item that is different.

CHAPTER

2 Sorting

theme
I Spy!

Use the Data

Name two objects that are the same in one way.

What You Will Learn

In this chapter you will learn how to:

- Sort objects.
- Use more and fewer to compare groups of objects.
- Solve problems by acting it out.

Dear Family,
In Chapter 2, I will learn to sort objects and compare groups. Here are some new vocabulary words and an activity that we can do together.

Button Hunt

- Draw a line down the middle of the paper to divide it in half. Have your child use the paper mat to sort the buttons into two groups. Discuss how the buttons in each group are the same.

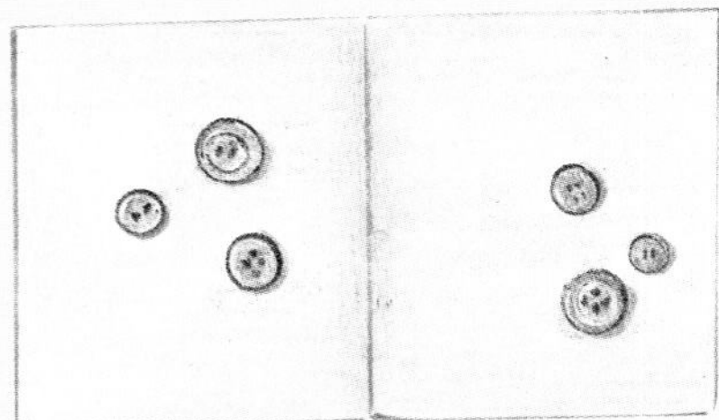

- Have your child sort the buttons in other ways. (size, color, number of holes, materials, etc.)

use

construction paper

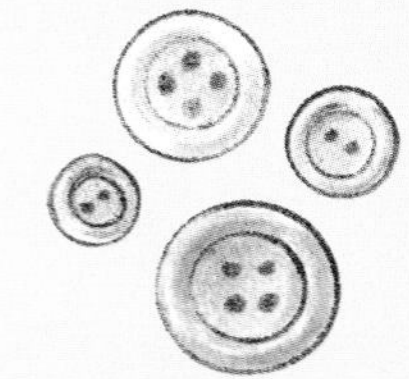

and buttons

or coins

Math Words

sort

To group objects that have things in common.

more

fewer

Additional activities at www.mhschool.com/math

Name____

Algebra & functions

Math Word
sort

Say this word.

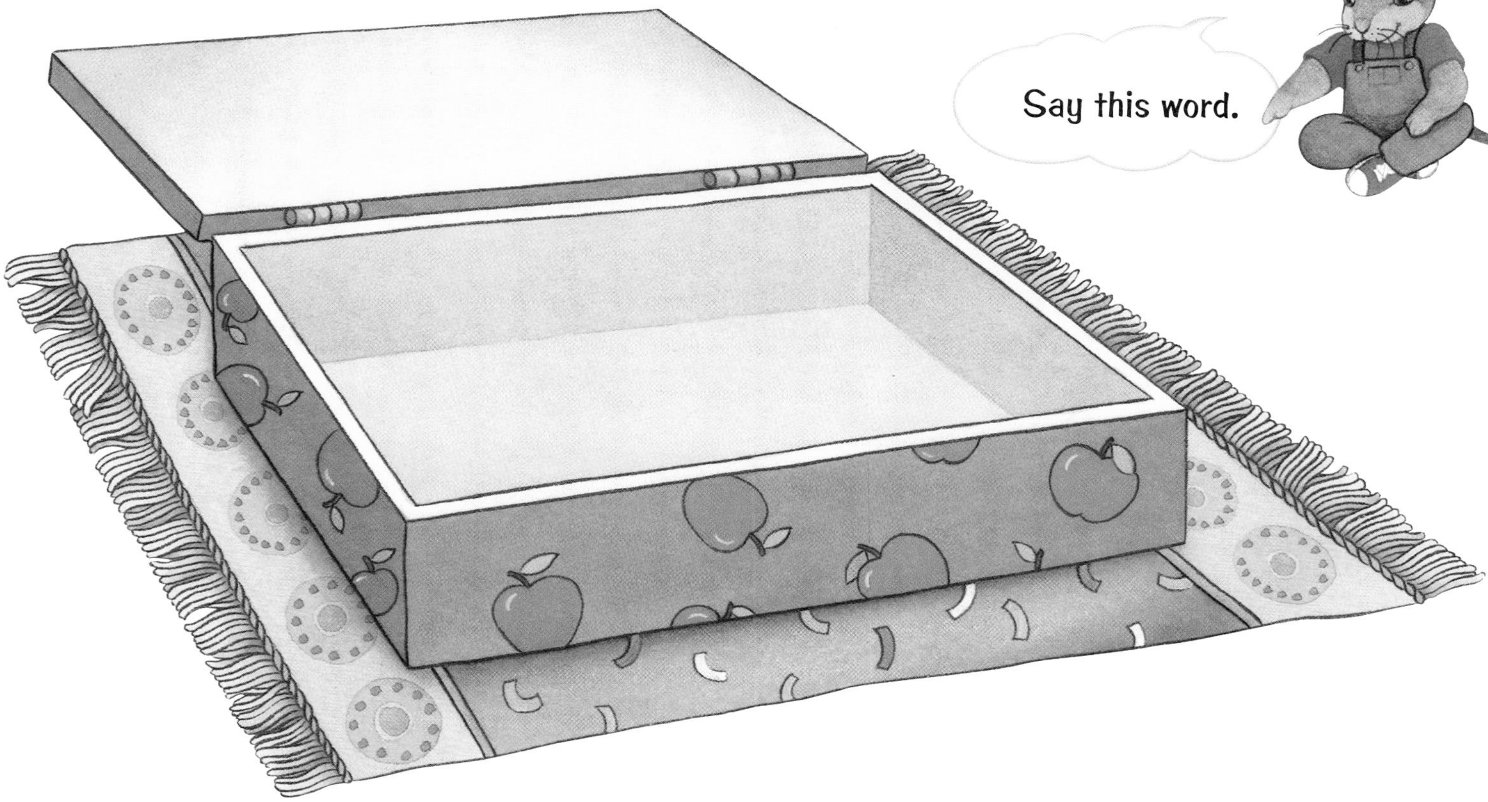

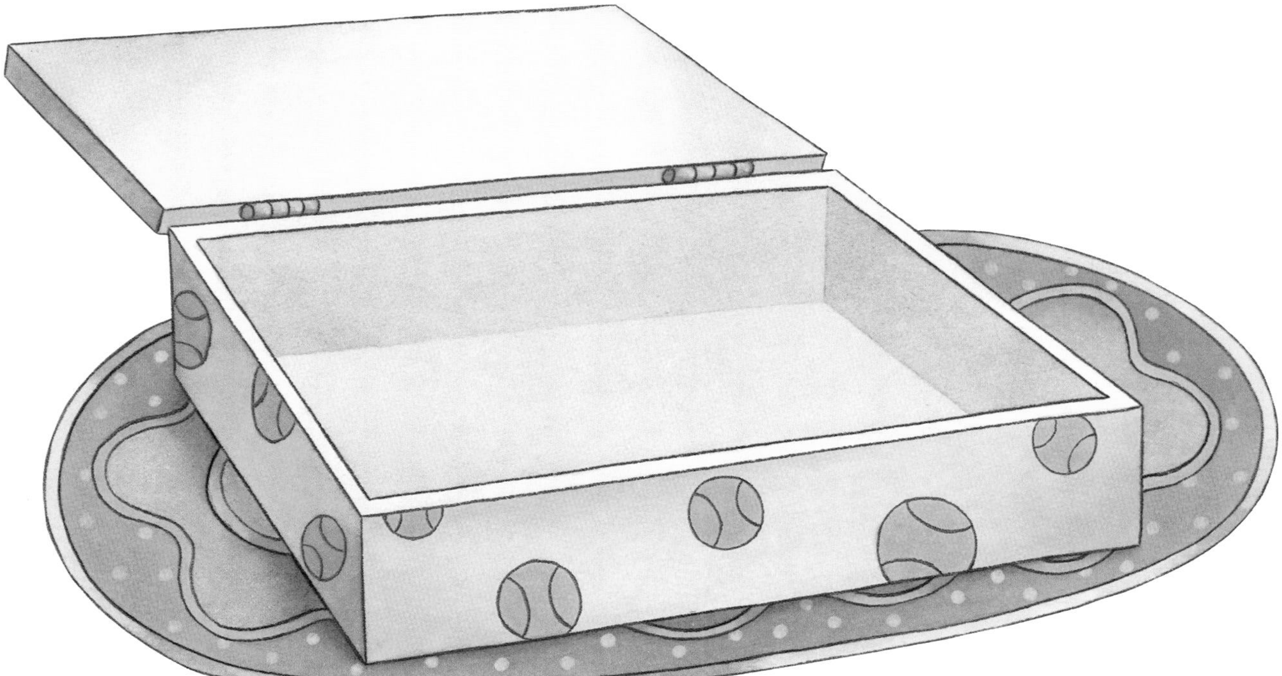

Use red and yellow cubes. Sort the cubes by color.
Draw the cubes in the toy chests to show 2 groups.

Math at Home: Your child sorted objects into different groups by color.
Activity: Cut out paper squares in two colors. Have your child sort the squares by color.

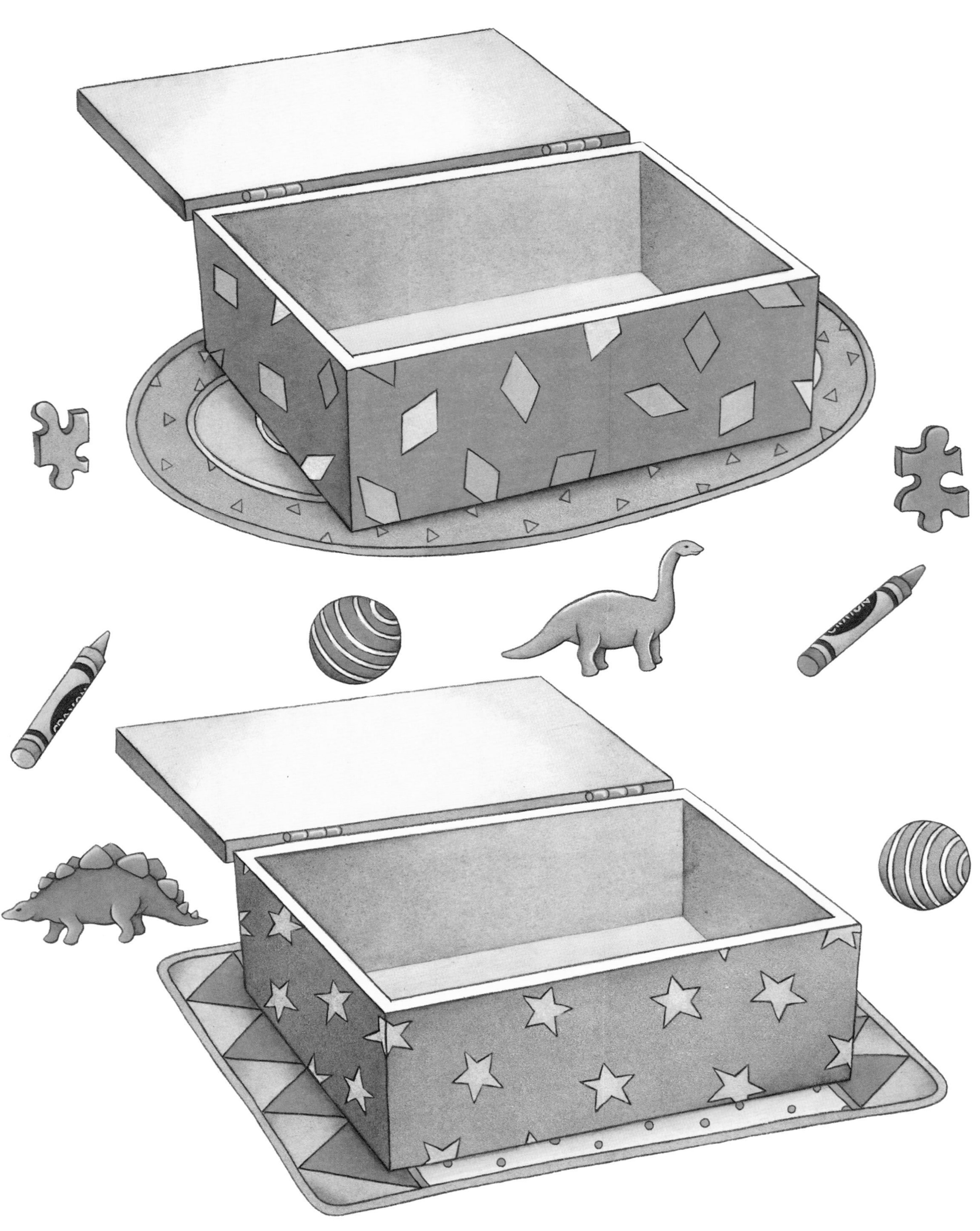

Draw a line from each object to the toy chest with the same color.

Name ______________________________

Algebra & functions

Draw lines from objects that are green and have wheels to the jar.
Draw something else in the jar that is green and has wheels.

Math at Home: Your child sorted objects by two attributes .
Activity: Ask your child to help you sort the family's socks by size and color.

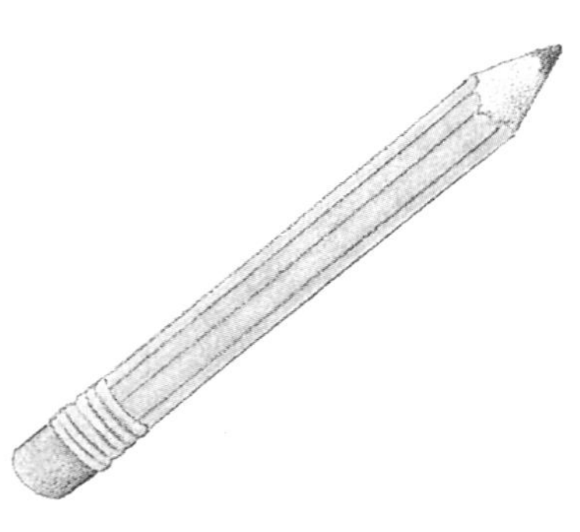

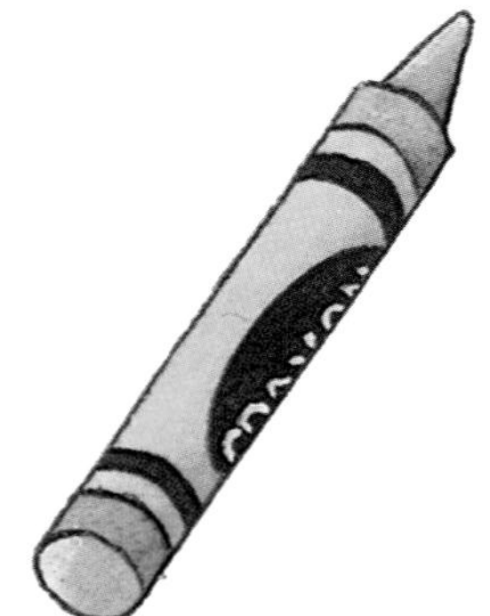

- Circle objects that are blue and round.
- Circle objects that are pink and you can use to write.
- Circle objects that are yellow things you can eat.

Name ______________________________

Algebra & functions

Color pictures that show animals orange.
Color pictures that show flowers green.

Math at Home: Your child sorted by coloring.
Activity: Show your child various foods such as fruits, crackers and pastas. Ask her or him to put the food into like groups.

Draw red lines from the large toy bucket to the large objects in the picture.
Draw blue lines from the small toy bucket to the smaller objects in the picture.

Name ______________________________

Act It Out

Use pattern blocks to make the shape. Color to show the pattern blocks you used.

Math at Home: Your child used pattern blocks to solve problems.
Activity: Ask your child to tell how he or she used pattern blocks to make the shape.

Use pattern blocks to make the shape. Color to show the pattern blocks you used.

Name ____________________

Math Words
more
fewer
same number

Use cubes to show more. Draw the cubes. ★ Use cubes to show fewer. Draw the cubes.
Use cubes to show same number. Draw the cubes.

Math at Home: Your child used connecting cubes and drew pictures to show more, fewer, and same number.
Activity: Make a group of four things, such as grapes or pencils. Ask your child to make one group with more objects, and then another group with fewer objects.

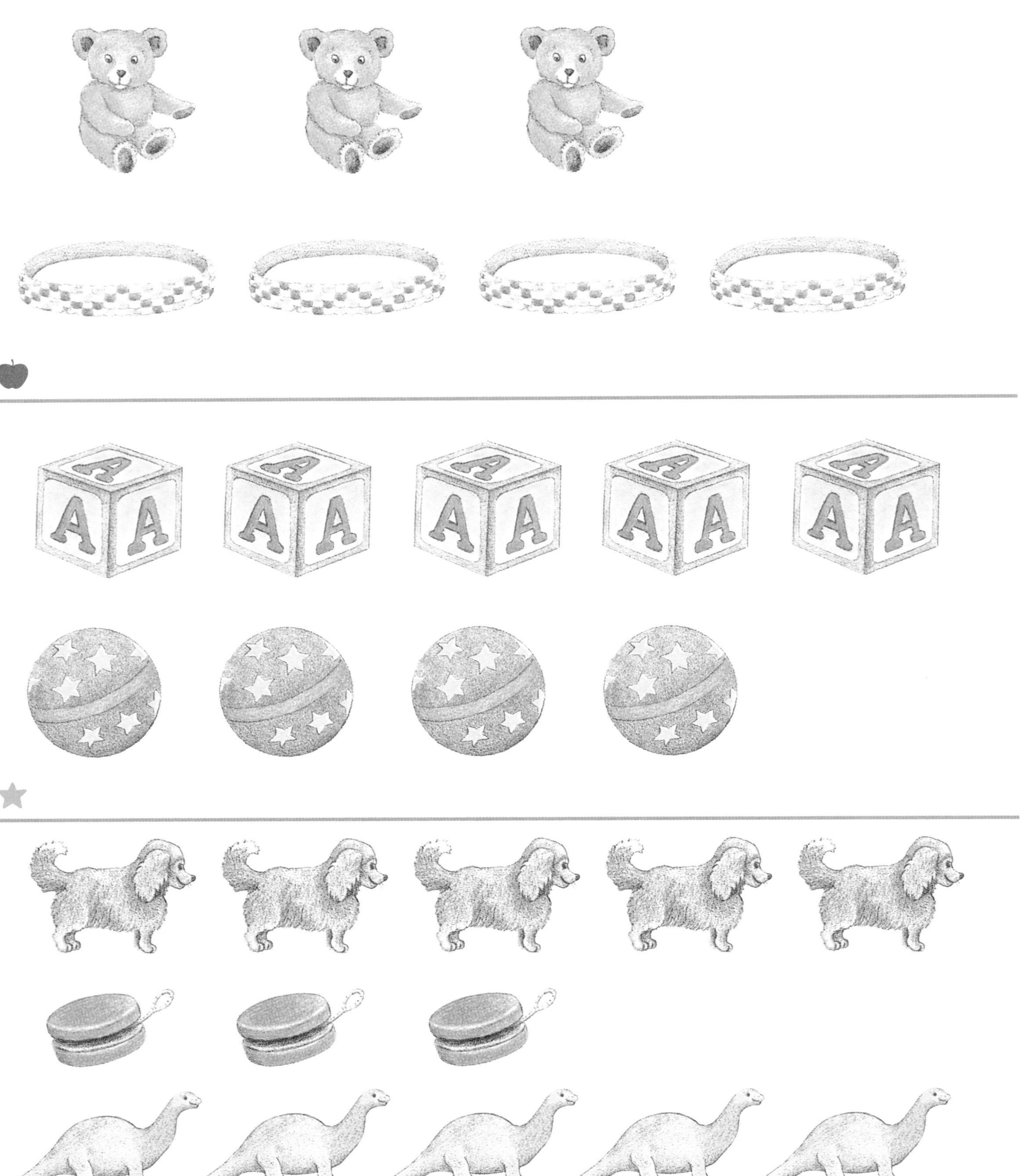

- Draw lines to match the collectibles. Circle the group that has more.
- Draw lines to match the collectibles. Circle the group that has fewer.
- Draw lines to match the collectibles. Circle the groups that have the same number.

Name ______________________________

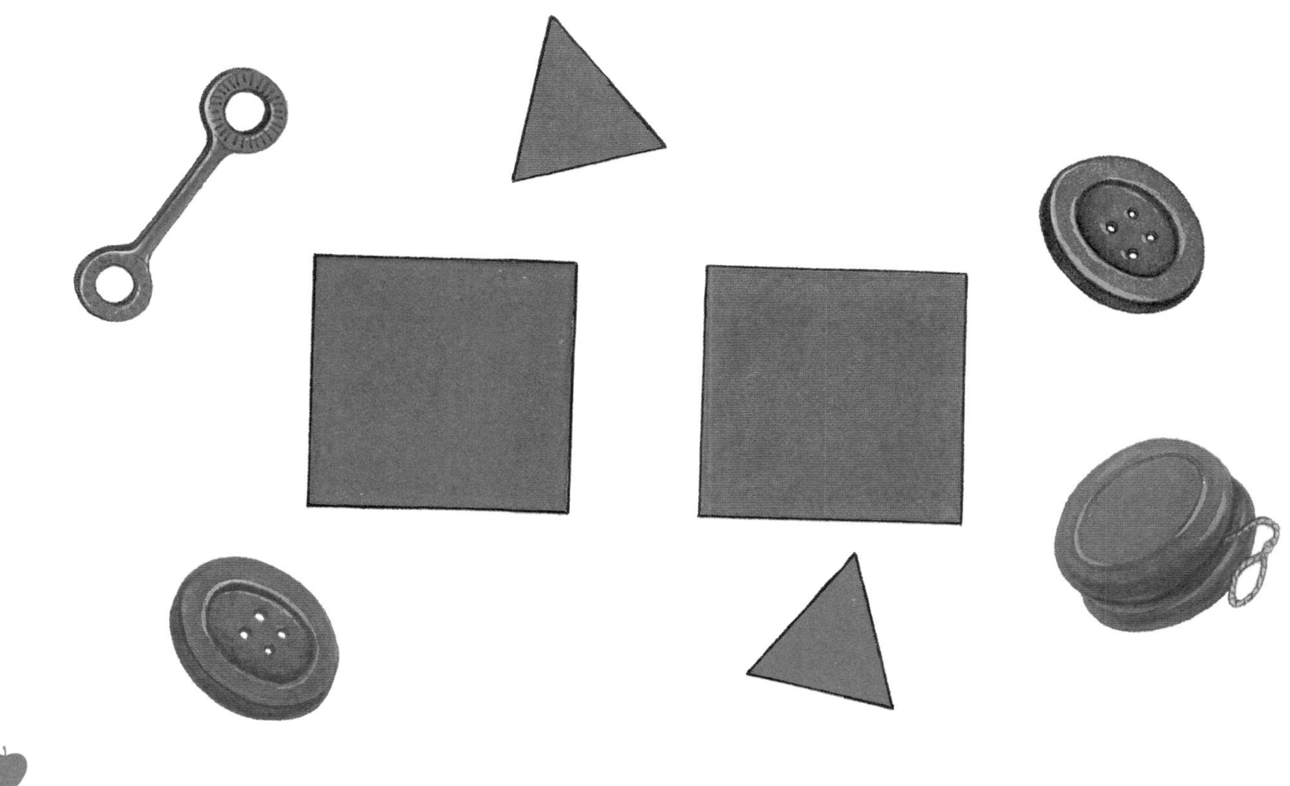

• Sort the objects by color. Draw a line from each object to the square of the same color.

★ Draw lines from the bugs that are small and red to the workspace. Draw another bug that is small and red.

- Draw lines to match the objects. Circle the group that has more.
- ★ Draw lines to match the objects. Circle the group that has fewer.
- Draw lines to match the objects. Circle the groups that have the same number.

CHAPTER 3

Data and Graphs

theme

Show and Tell

Use the Data

What different types of instruments do you see?

What You Will Learn

In this chapter you will learn how to:

- Make real graphs, picture graphs and bar graphs.
- Solve problems by making a real graph.

Dear Family,
In Chapter 3, I will learn to make picture and bar graphs. Here are some new vocabulary words and an activity that we can do together.

Comparing Coins

- Draw a line down the center of a piece of paper to divide it into two equal columns.Have your child take a handful of coins and sort them on the paper.

- Have your child put the coins in two columns by type, and then tell you which column has more. You can repeat the activity with a different handful of coins.

use

nickels and pennies

paper

Math Words

graph
A graph is used to show information.

real graph
A graph using real objects.

picture graph
A graph using pictures.

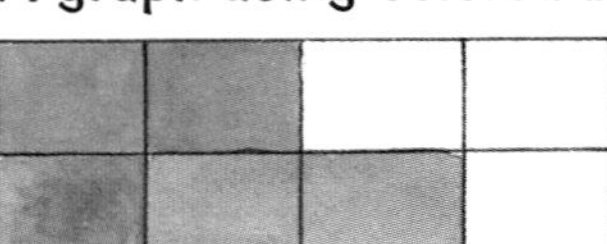

bar graph
A graph using colored bars.

Additional activities at www.mhschool.com/math

Name______________________________

Look at the books and bears at the top of the page. Color a picture on the graph to show each.

Math at Home: Your child used information to make a picture graph.
Activity: Have your child tell you which row has more objects colored.

Which is the favorite pet—a cat or a dog? Ask 5 people.
Color a picture on the graph to show each choice.

Which animal do more children like? Circle the answer.

Name ____________________

Make a Real Graph

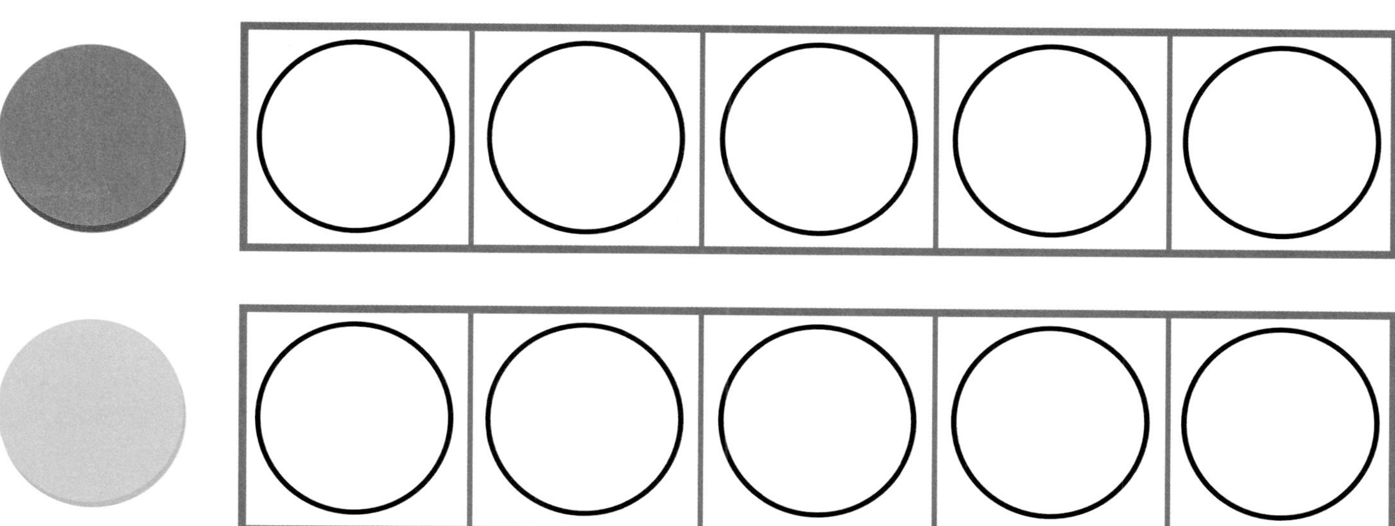

Grab a handful of 2-color counters. Sort them on the mat and then put them on the graph. Color the graph to show the counters. Are there more ● or ● ? Circle the answer.

Math at Home: Your child made a real graph with colored counters to solve problems.
Activity: Have your child make a real graph with two different colors of socks.

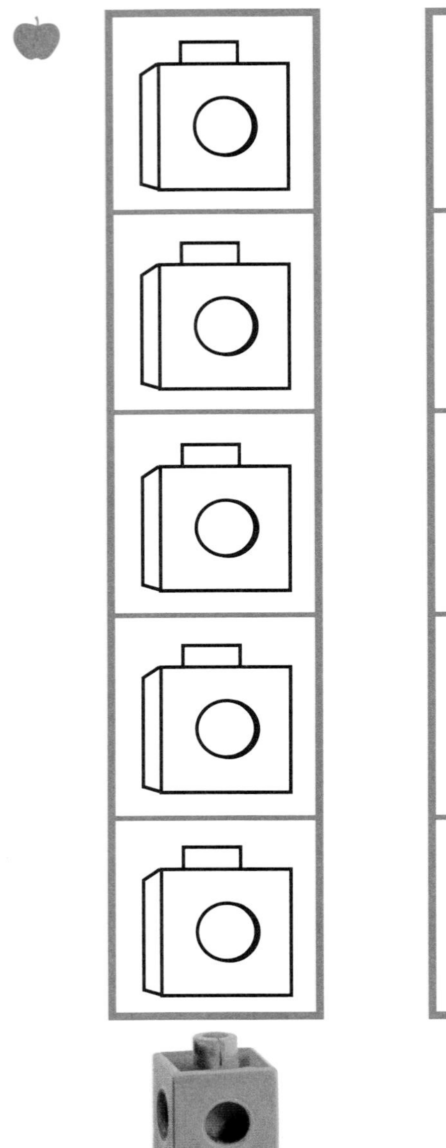

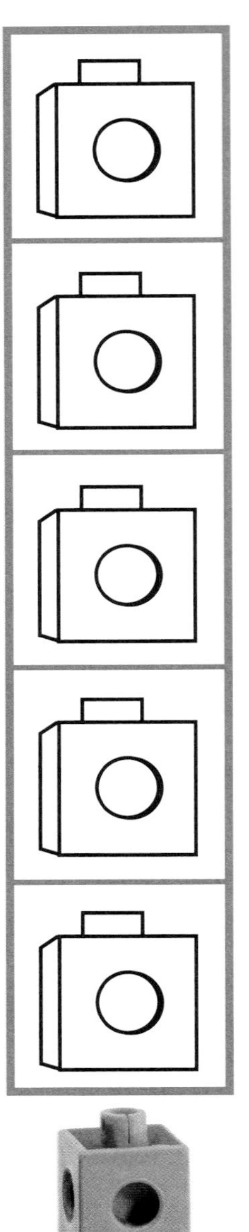

Grab a handful of red and green cubes. Sort them on the mat and then put them on the graph. Color the graph to show the cubes.

★ Are there fewer or ? Circle the answer.

Name ____________________

Say these words.

Math Word
bar graph

	6
	5
4	4
3	3
2	2
1	1

more

fewer

Look at the apples in the tree. Color a box on the graph to show each apple.
Are there more red apples or green apples? Draw pictures to show more and fewer.

Math at Home: Your child made a bar graph to compare the number of red apples and green apples.
Activity: Have your child make a bar graph to show which color apple your family likes better.

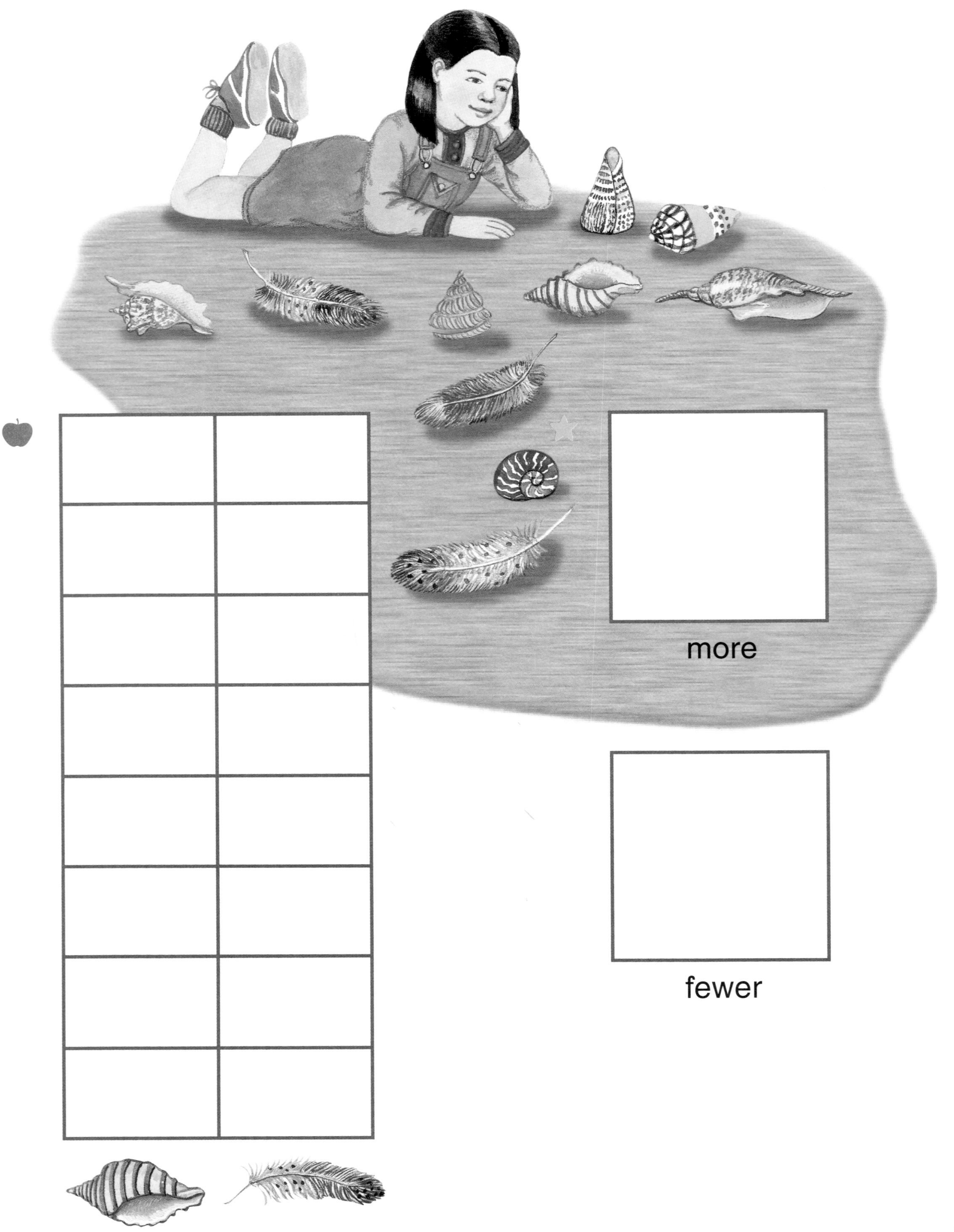

- Look at the items on the rug. Color a box in the graph to show each item. Are there more feathers or seashells?
- ★ Draw pictures to show more and fewer.

Name ____

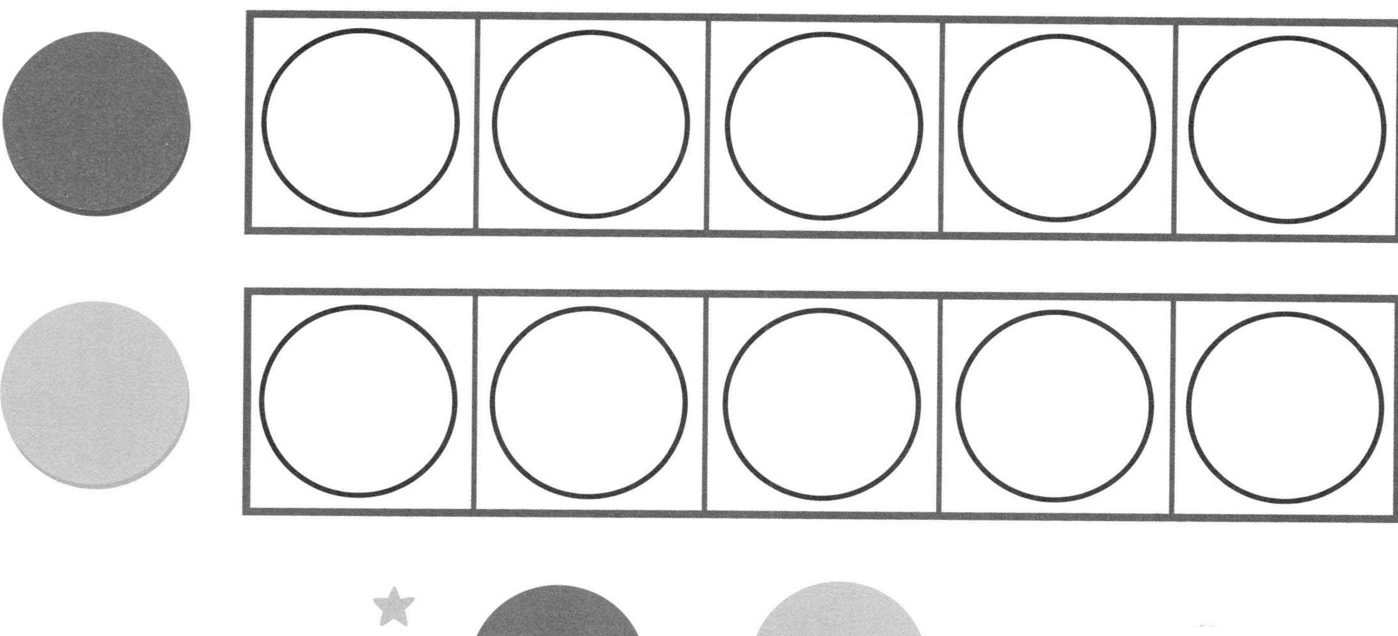

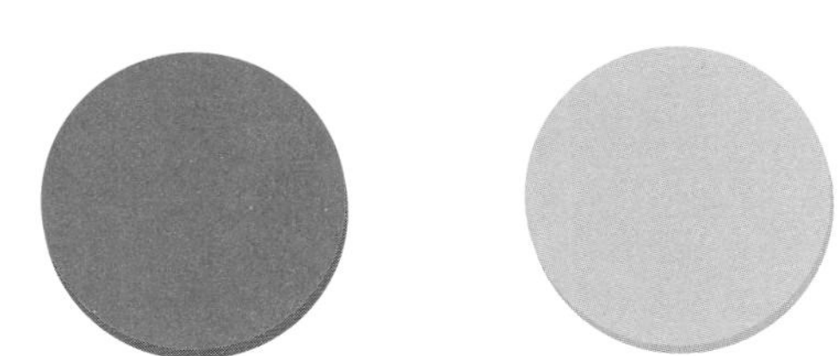

- Look at the counters at the top of the page. Color a picture on the graph to show each counter.
- ★ Are there more red or yellow counters? Circle the answer.

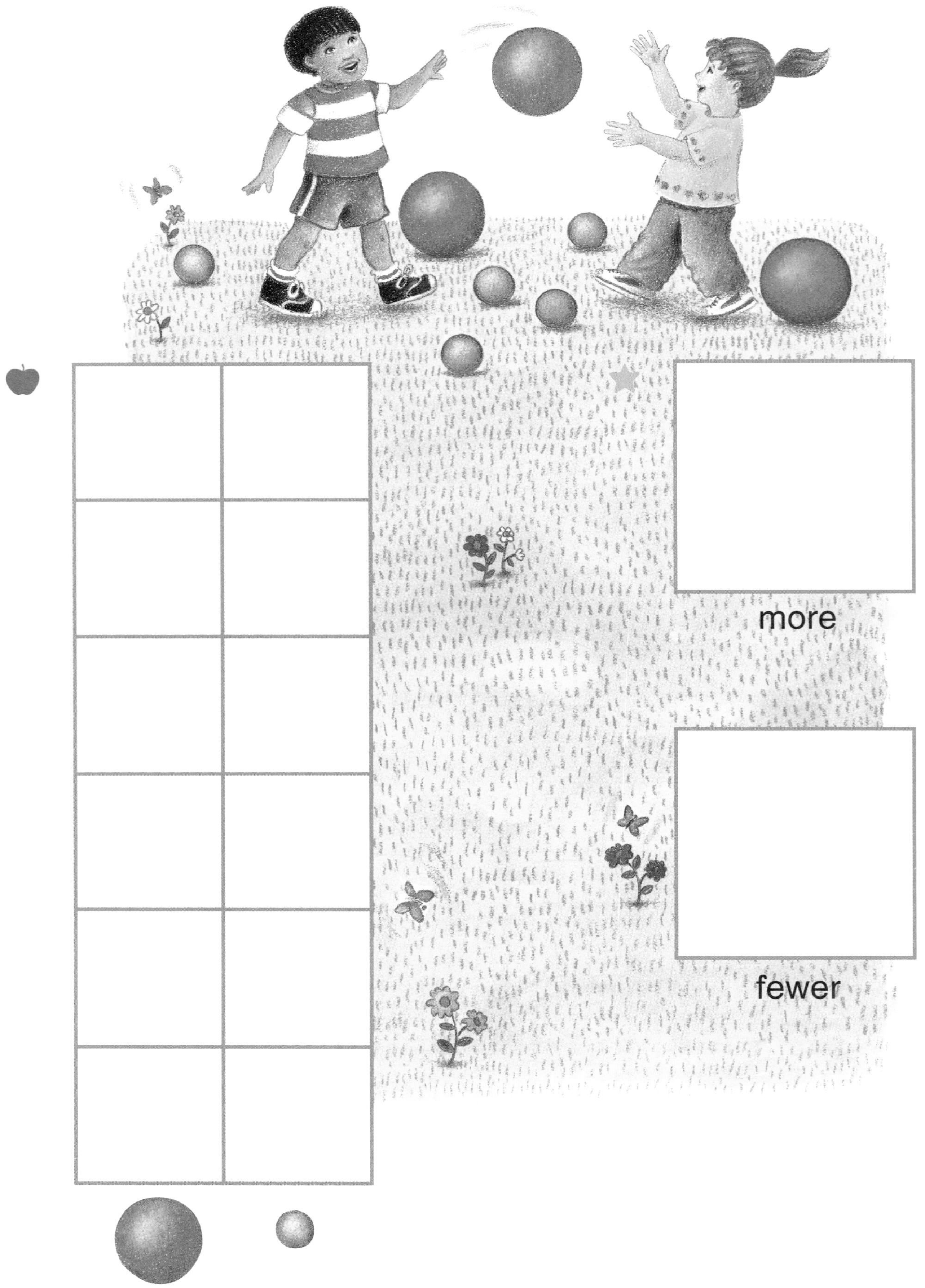

- Look at the balls. Color a box in the graph to show each ball.
- Draw pictures to show more and fewer.

CHAPTER 4

Patterns

theme

Going on a Safari

Use the Data

What patterns do you notice?

What You Will Learn

In this chapter you will learn how to:

- Copy patterns.
- Extend patterns.
- Find and use patterns to solve problems.

Dear Family,
In Chapter 4, I will learn to make simple patterns. Here are some new vocabulary words and an activity that we can do together.

Math Words

pattern

circle

square 

triangle 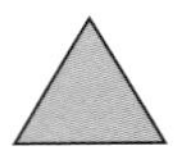

rectangle

Nickel and Penny Patterns

- Make the following pattern: *nickel, penny, nickel, penny, nickel, penny.*

use

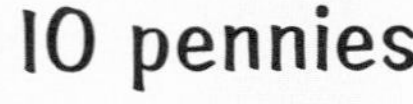

- Have your child name each coin. Ask your child what coin could come next in the pattern.
- Repeat the activity using the pattern: *nickel, nickel, penny, penny, nickel, nickel, penny, penny.*

Additional activities at www.mhschool.com/math

Name ______________________________

Math Word
pattern

Copy each pattern. Circle the action that could come next. Tell how you know.

Math at Home: Your child is learning about simple patterns.
Activity: Help your child find simple patterns in wallpaper, wrapping paper, or in fabric, at home or while shopping.

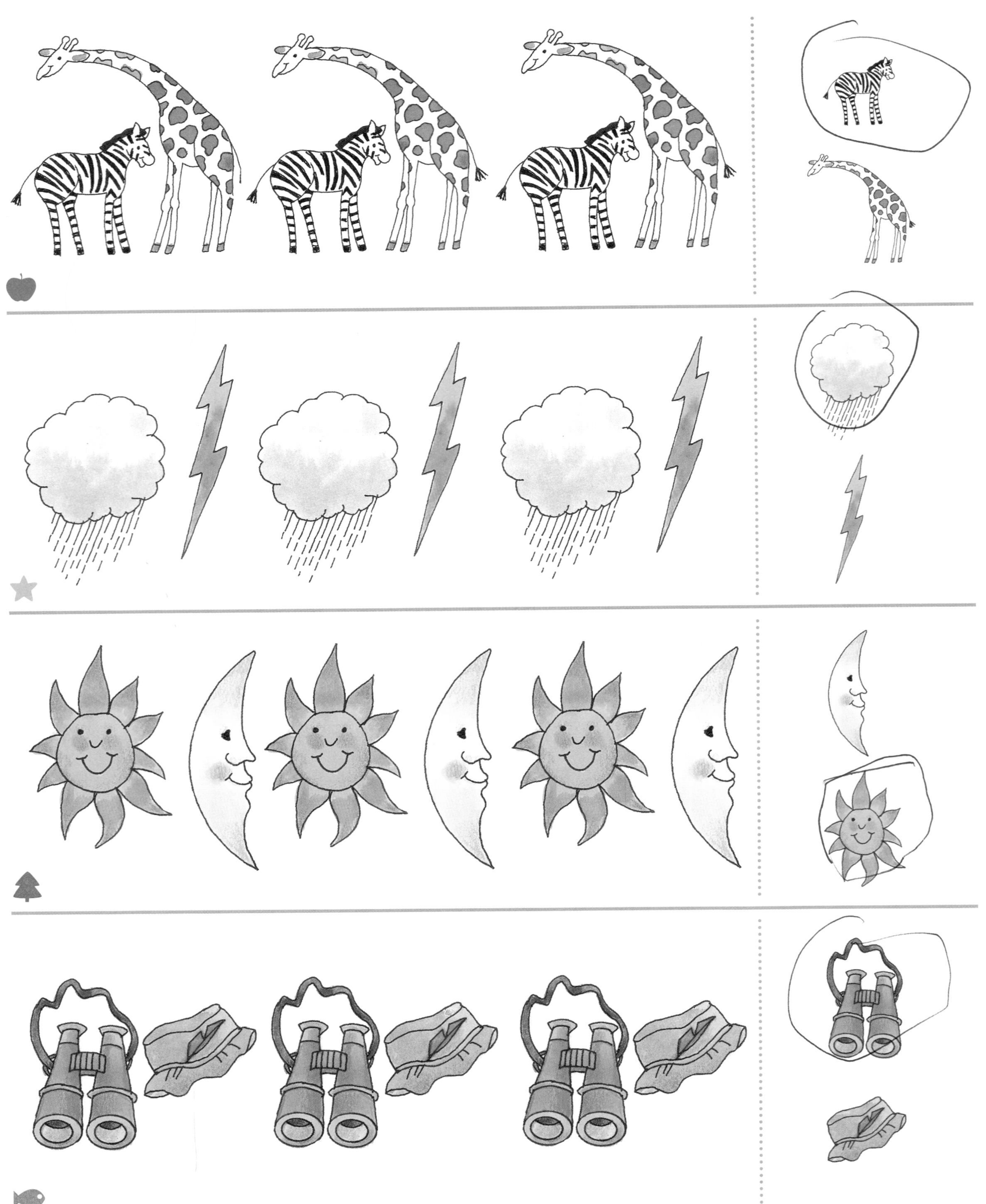

Look at each pattern. Circle what could come next in the pattern. Tell how you know.

Name ____________________

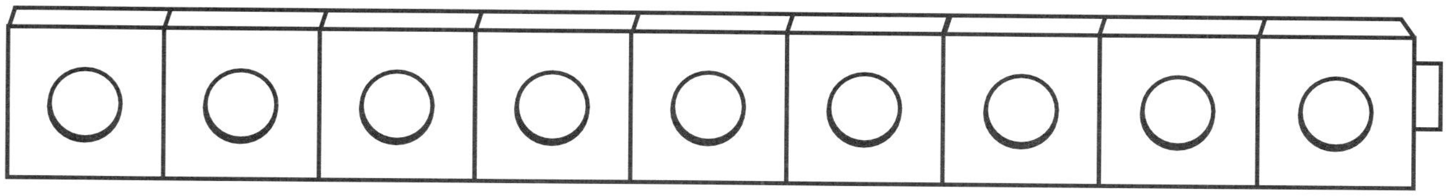

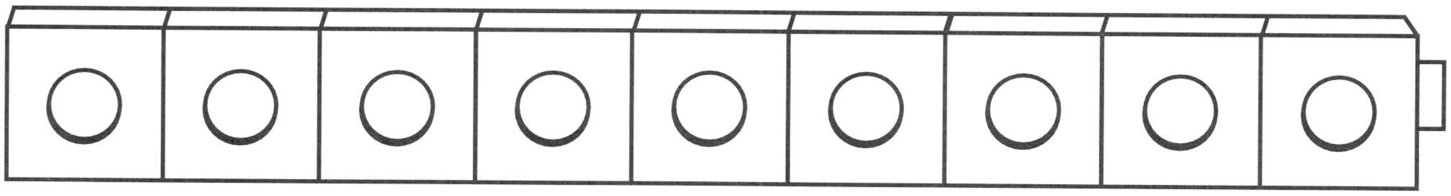

Color the cubes to copy the pattern.

Math at Home: Your child copied two color patterns.
Activity: Draw ten boxes in a row. Ask your child to color the boxes to show a pattern with yellow and purple.

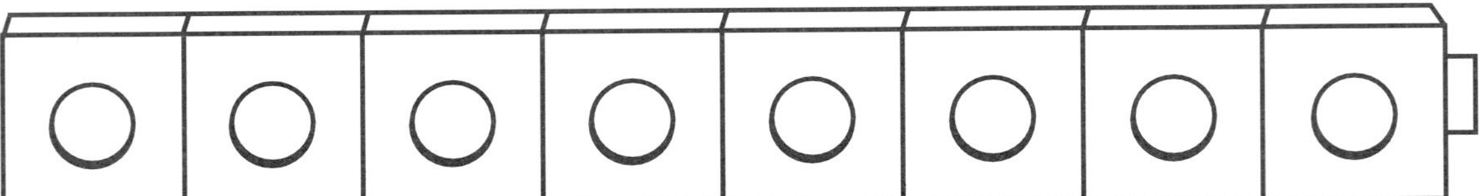

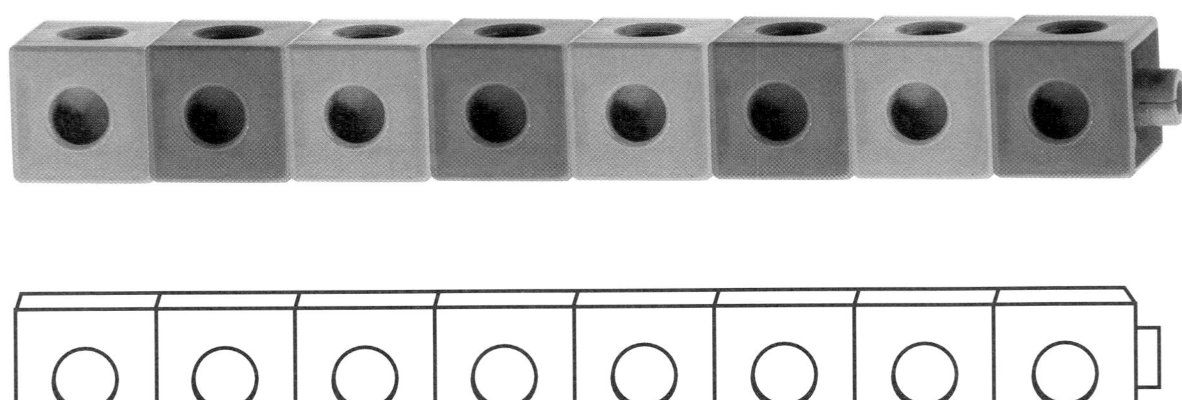

Color the cubes to copy the pattern.

Name ______________________________

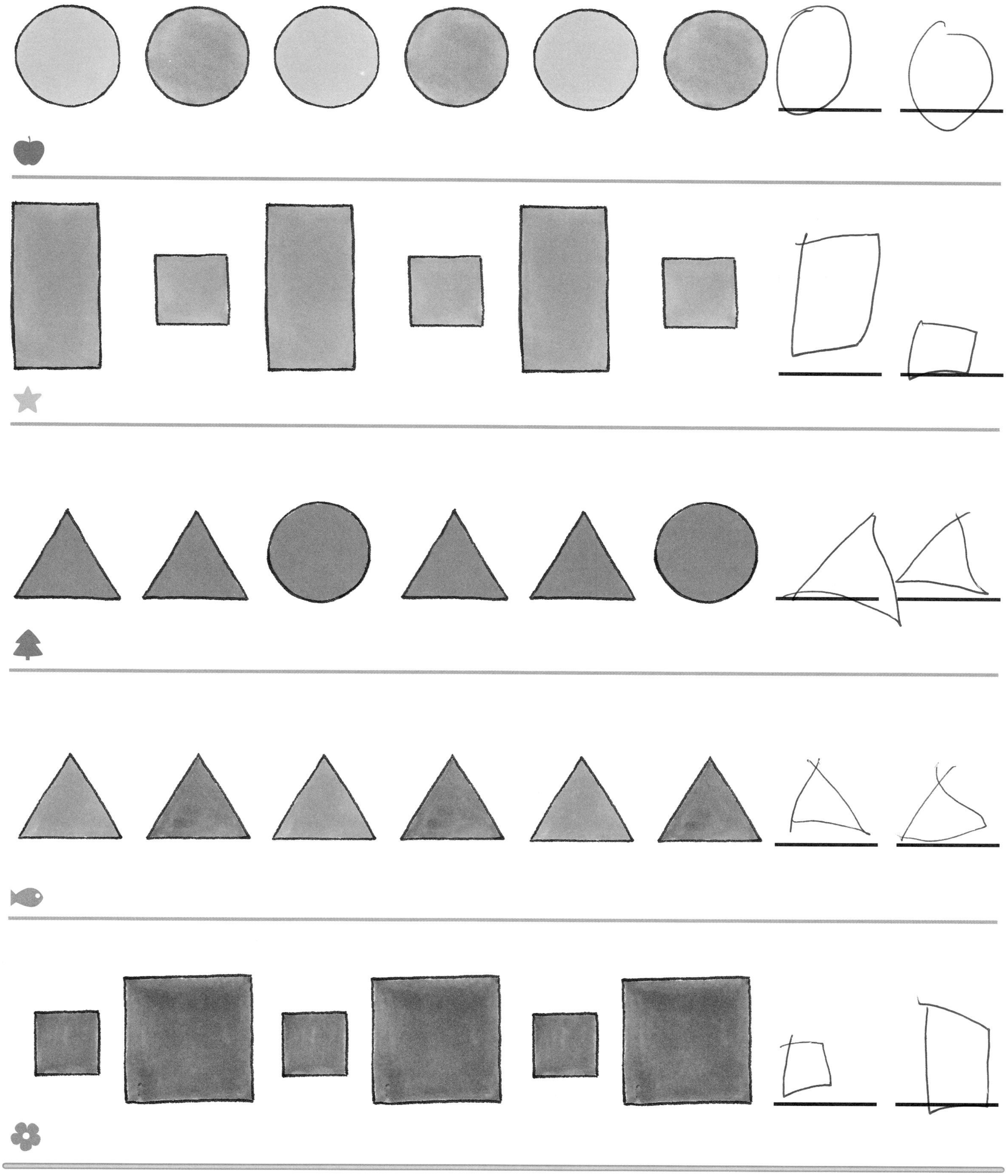

Draw shapes to show what could come next in the pattern. Tell how you know.

Math at Home: Your child is learning to extend patterns.
Activity: Draw a pattern of alternating red circles and red squares across a page. Have your child draw shapes to extend the pattern.

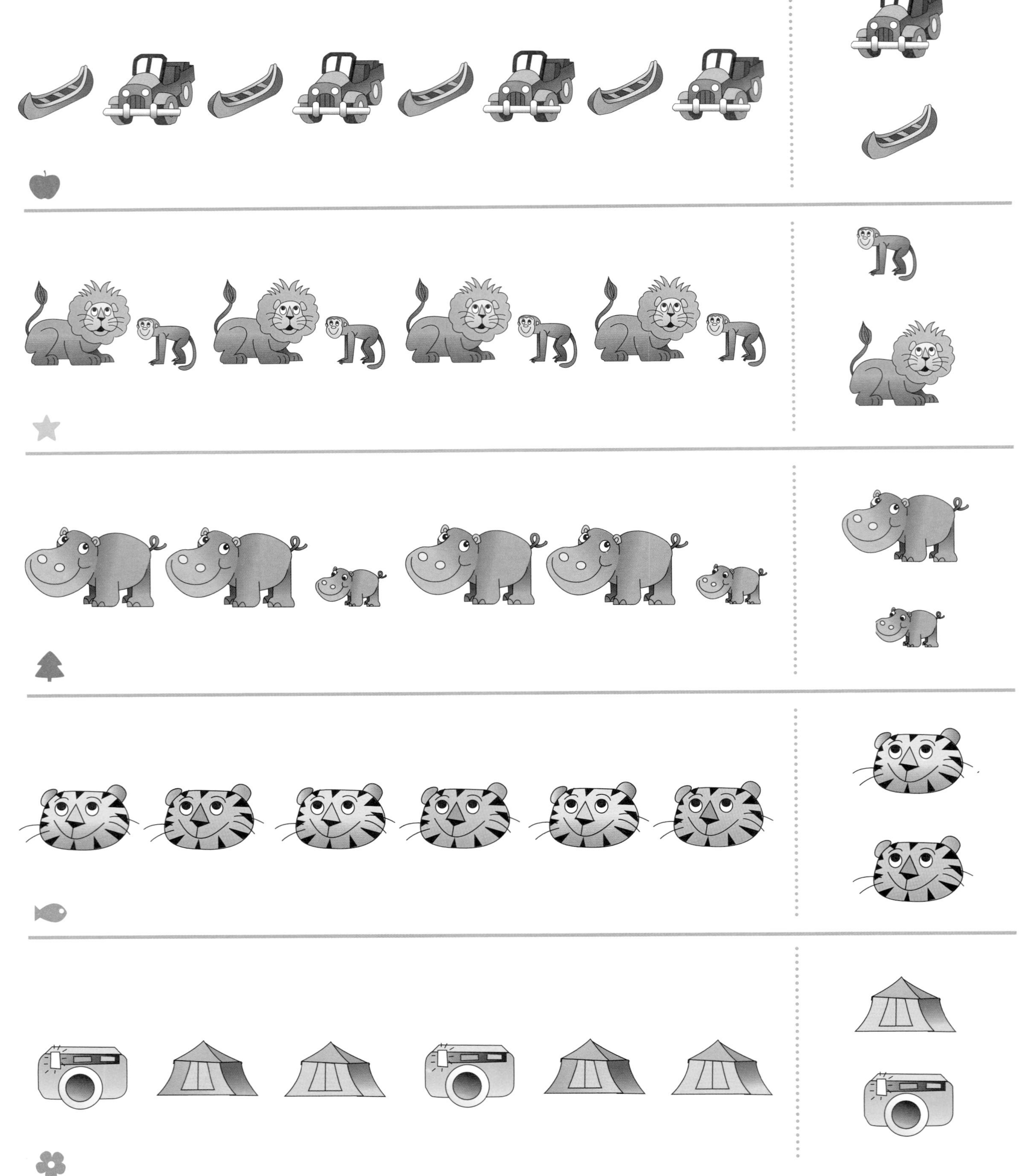

Circle what could come next in the pattern. Tell how you know.

Name ______________________________

Find a Pattern

Color to finish the pattern.

Math at Home: Your child is learning to extend patterns.
Activity: Draw a pattern of a red square, a blue square, a red square, and a blue square. Ask your child to draw three more squares to continue the pattern.

Color to finish the pattern.

Name ______________________________

Complete a Pattern

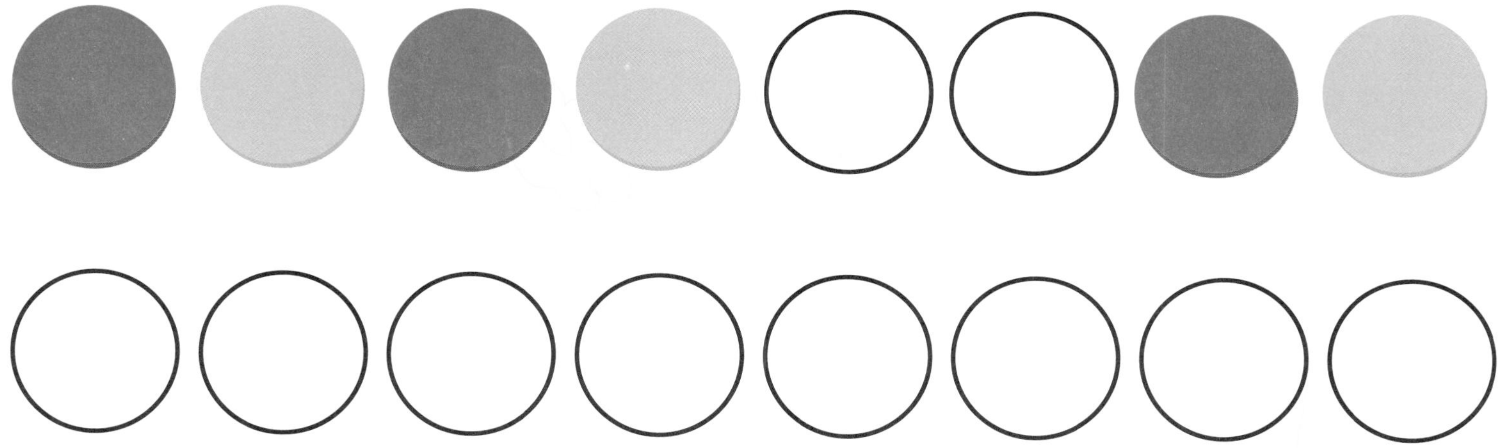

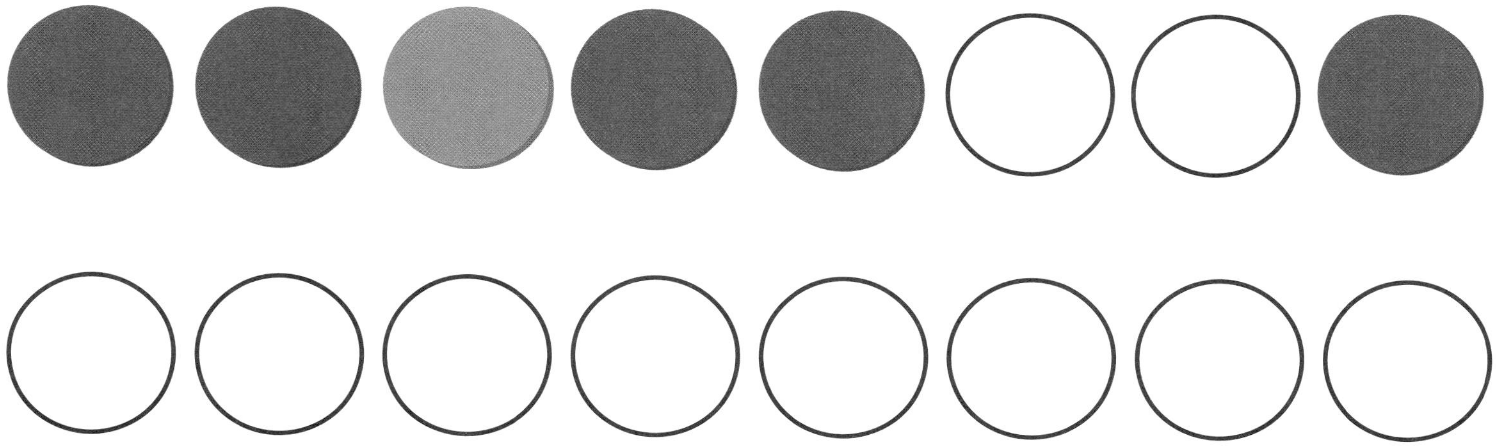

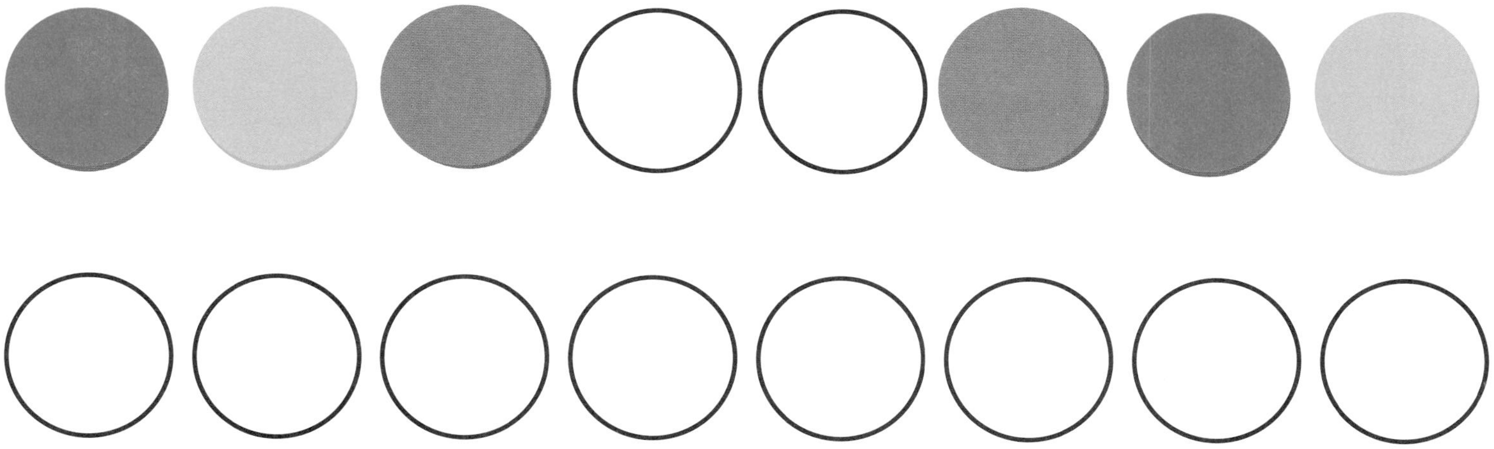

What could the missing counters be?
Use counters to copy and complete each pattern. Color the counters.

Math at Home: Your child used counters to complete a pattern.
Activity: Have your child make a pattern of forks and spoons. Take one object away and have your child tell you what is missing from the pattern.

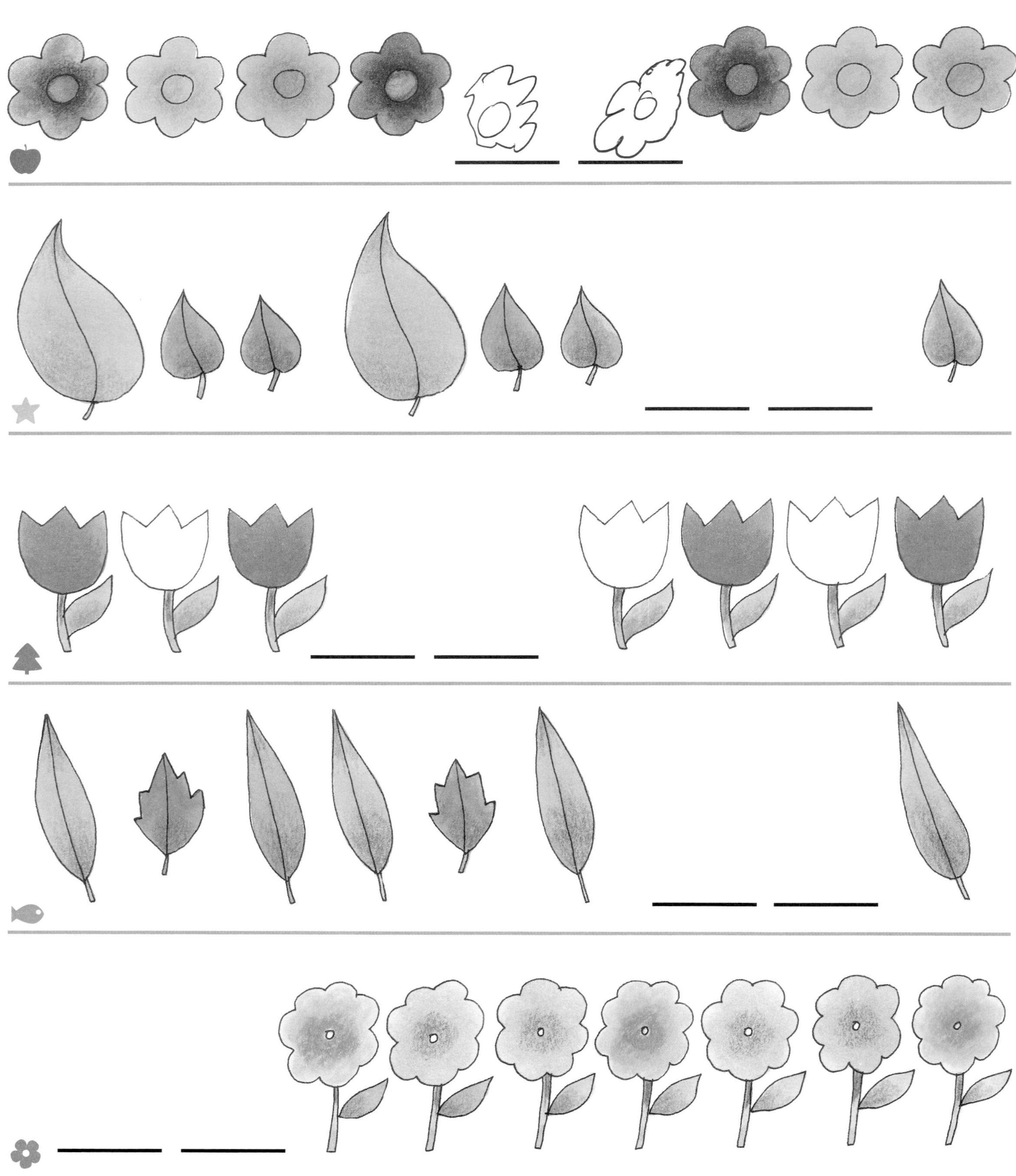

Draw and color to complete each pattern.

Name ______________________________

Color the cubes to copy the pattern.

Circle what could come next in each pattern.

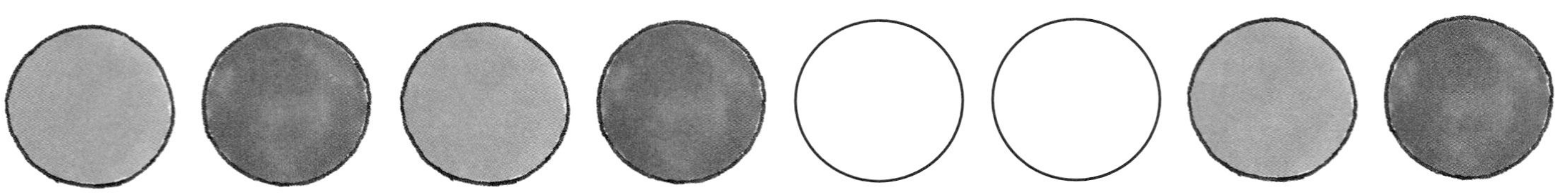

★

● ★ Color to complete each pattern.

▲ Color to finish the pattern.

CHAPTER 5

Numbers to 5

theme
Out on the Farm

Use the Data

How many piglets do you see?

What You Will Learn

In this chapter you will learn how to:

- Count, read and write numbers to 5.
- Compare and order numbers to 5.
- Make picture graphs to solve problems.

Dear Family,
In Chapter 5, I will learn to count, read and write numbers to 5. Here are some new vocabulary words and an activity that we can do together.

Let's Count

Math Words

count

1 2 3

Count to find how many.

equal to

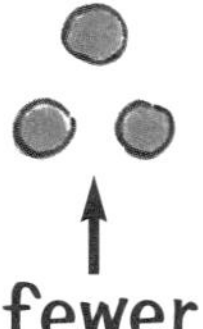 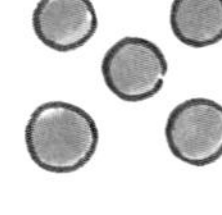

fewer

 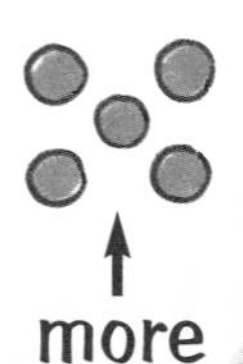

more

- Number the plates 0 to 5.

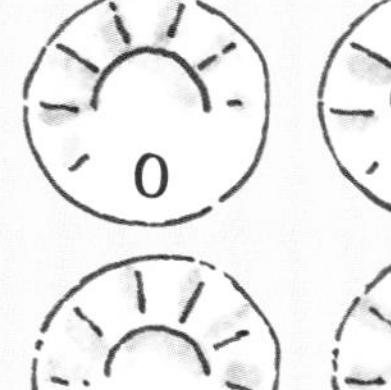 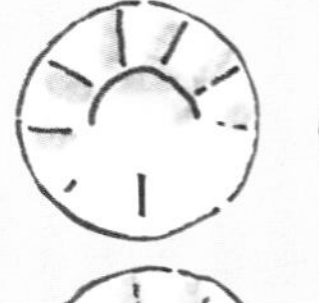  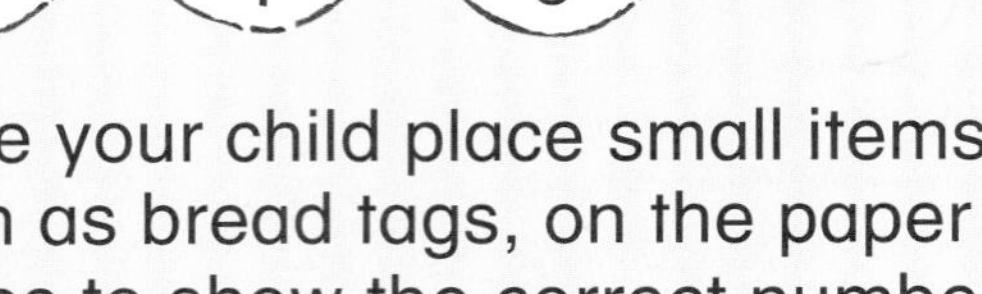

- Have your child place small items, such as bread tags, on the paper plates to show the correct number.
- Then ask your child to check by counting the number of objects in each group again.

use

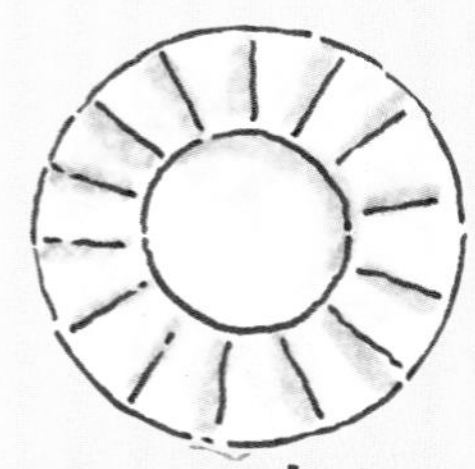

paper plates

small toys or

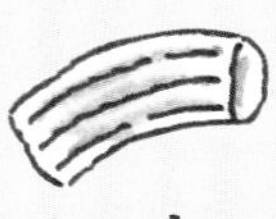

pasta

Additional activities at www.mhschool.com/math

Name ____________________

1

2

3

Use cubes to show 1, 2, and 3. Draw the cubes. Write the number.

Math at Home: Your child is learning to count, read, and write 1, 2, and 3.
Activity: Write numerals 1, 2, and 3 on separate index cards. Place small household objects in groups of 1, 2, and 3, and ask your child to hold up the card that tells the number in each group.

Count the animals. Write the number.

Name ____________________

4
four

5
five

4

5

Use cubes to show 4 and 5. Draw the cubes. Write the number.

Math at Home: Your child is learning to count, read and write the numbers 4 and 5.
Activity: Have your child find groups of 4 and 5 objects and tell how many are in each group.

Count the animals. Write the numbers.

Name ____________________

Use cubes to show the number 3. Draw the cubes. Take the cubes away to show 0. Write the numbers.

Math at Home: Your child learned about the meaning of 0 and wrote the number.
Activity: Have your child count out 3 small items. Take them away one at a time, and recount the group. When all the items are removed, have your child tell how many are left.

Count how many animals. Write the number. Circle each group that shows zero.

Name ______________________________

2 3 4

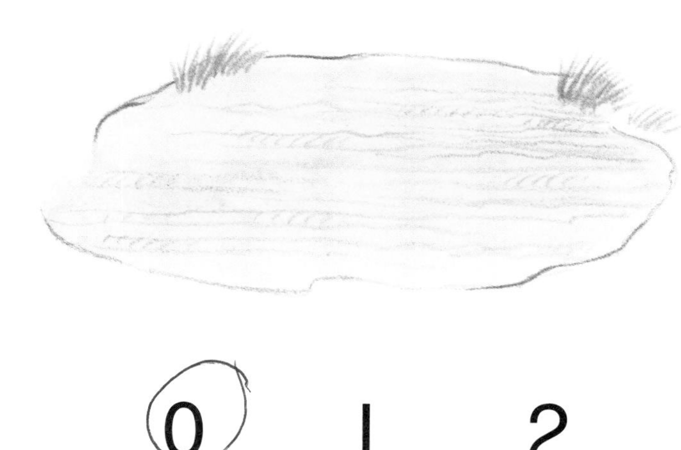

0 1 2

1 2 3

0 1 2

3 4 5

1 2 3

Count the animals. Circle the number.

Math at Home: Your child practiced numbers to 5.
Activity: Have your child find groups of 1 to 5 objects in your house and tell how many.

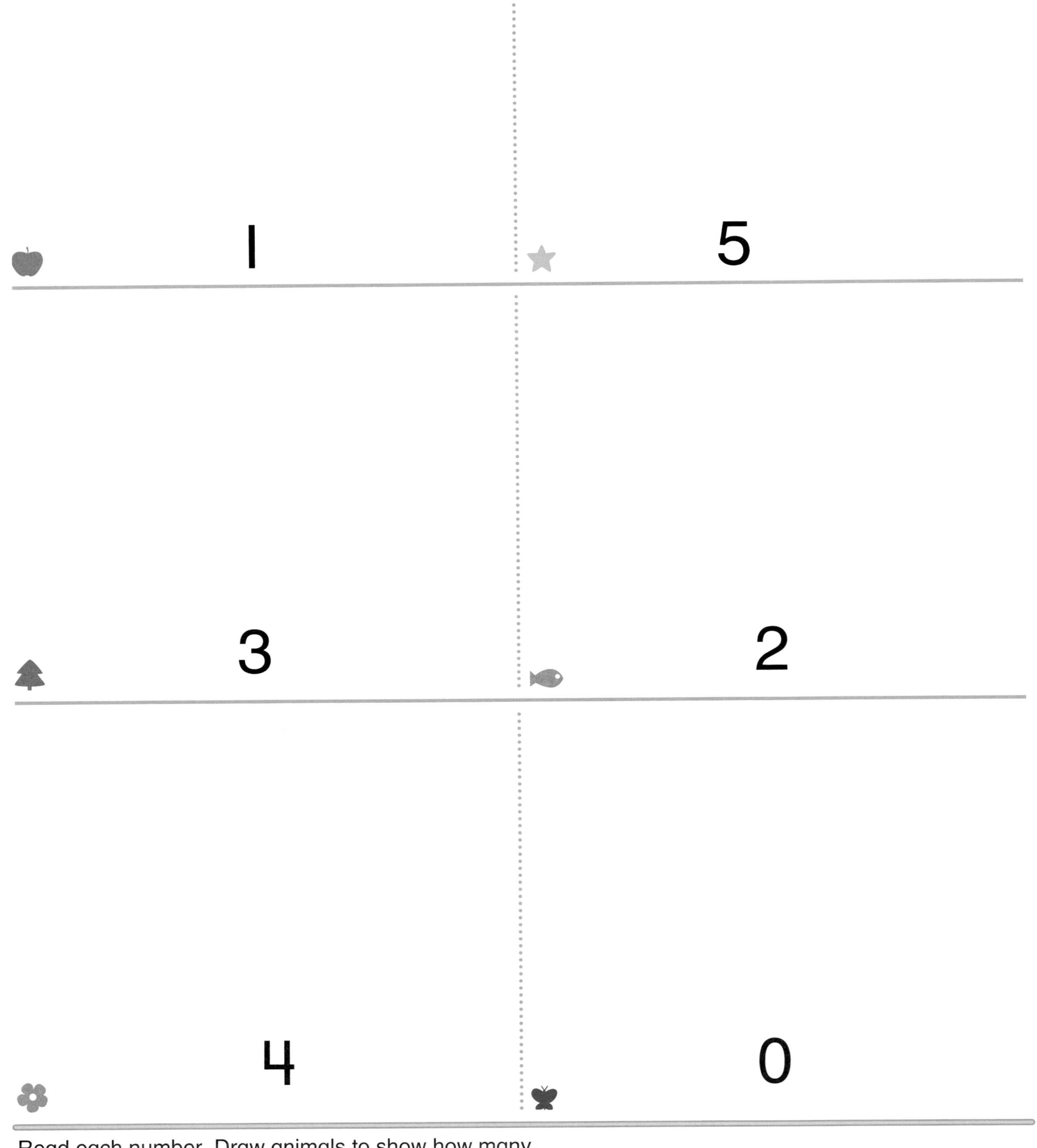

Read each number. Draw animals to show how many.

Name ______________________________

Make a Graph

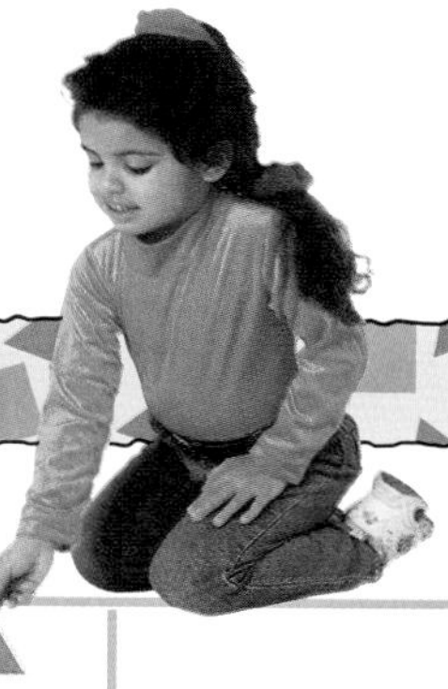

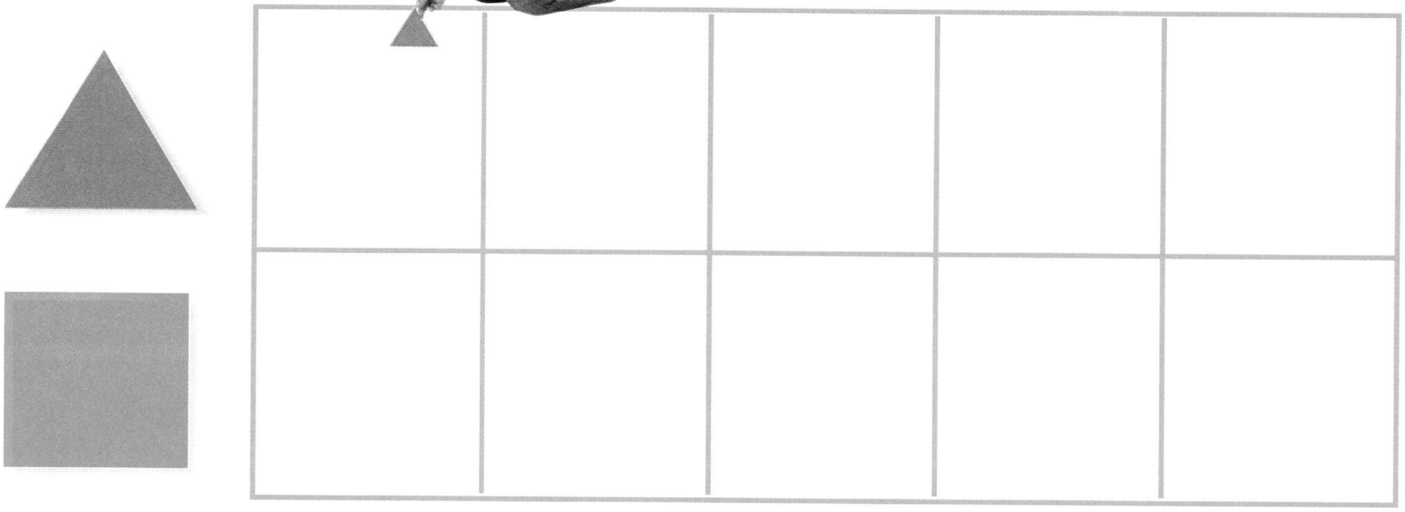

Take five pattern blocks. Put them on the graph. Draw the pattern blocks. Count. Write how many.

Math at Home: Your child solved problems by using pattern blocks and making a graph.
Activity: Have your child draw a picture of 5 rings and 2 boxes. Then help your child make a picture graph to show the groups and write the numbers.

Problem Solving • Strategy

- Color a picture on the graph to show each animal.
- Count the animals. Write the number.

Name ____________________

    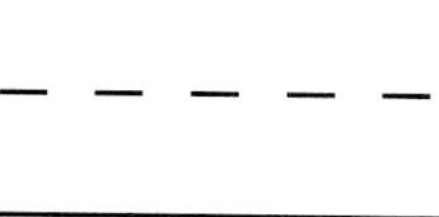

   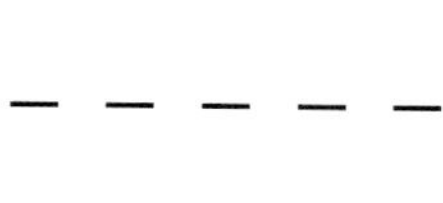

   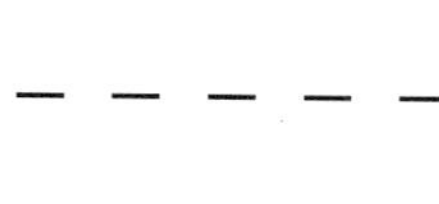

   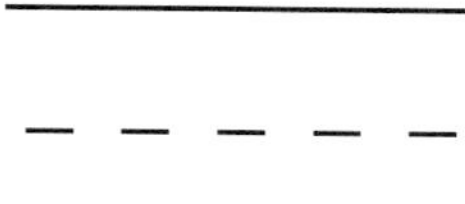

  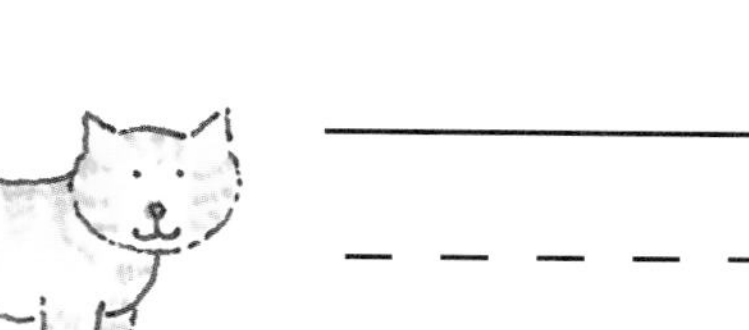

Count each group of animals. Write the numbers. Draw lines to match the animals. Circle groups that have the same number.

Math at Home: Your child counted objects and drew circles around groups that had the same numbers.
Activity: Make three groups of items, such as paper clips. Place 3 items in each of two groups and 5 items in the third group. Ask your child to find the two groups that are equal.

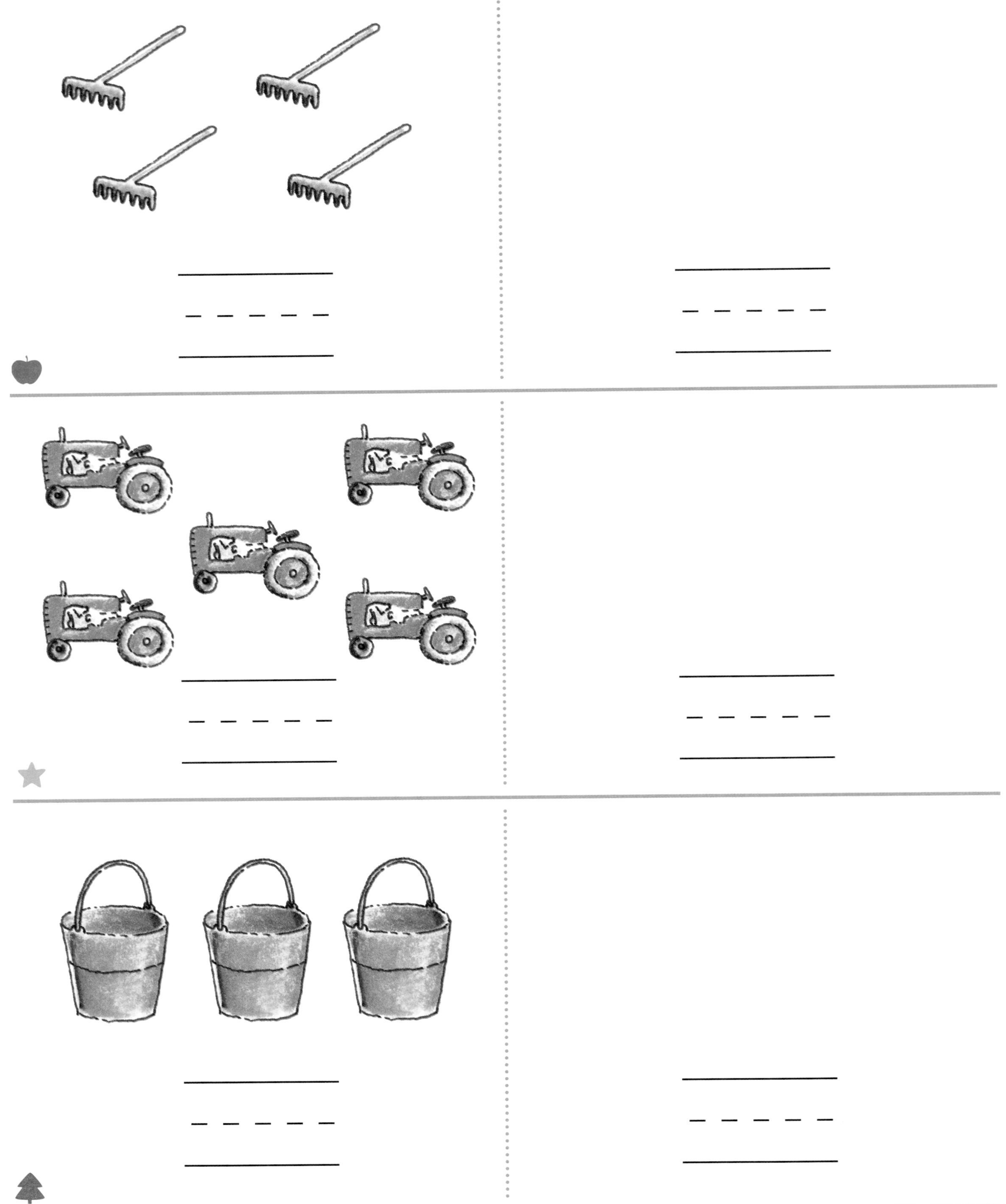

Count each group. Write the number. Draw a group that is equal to the first group.

Name ______________________________

Compare Numbers to 5

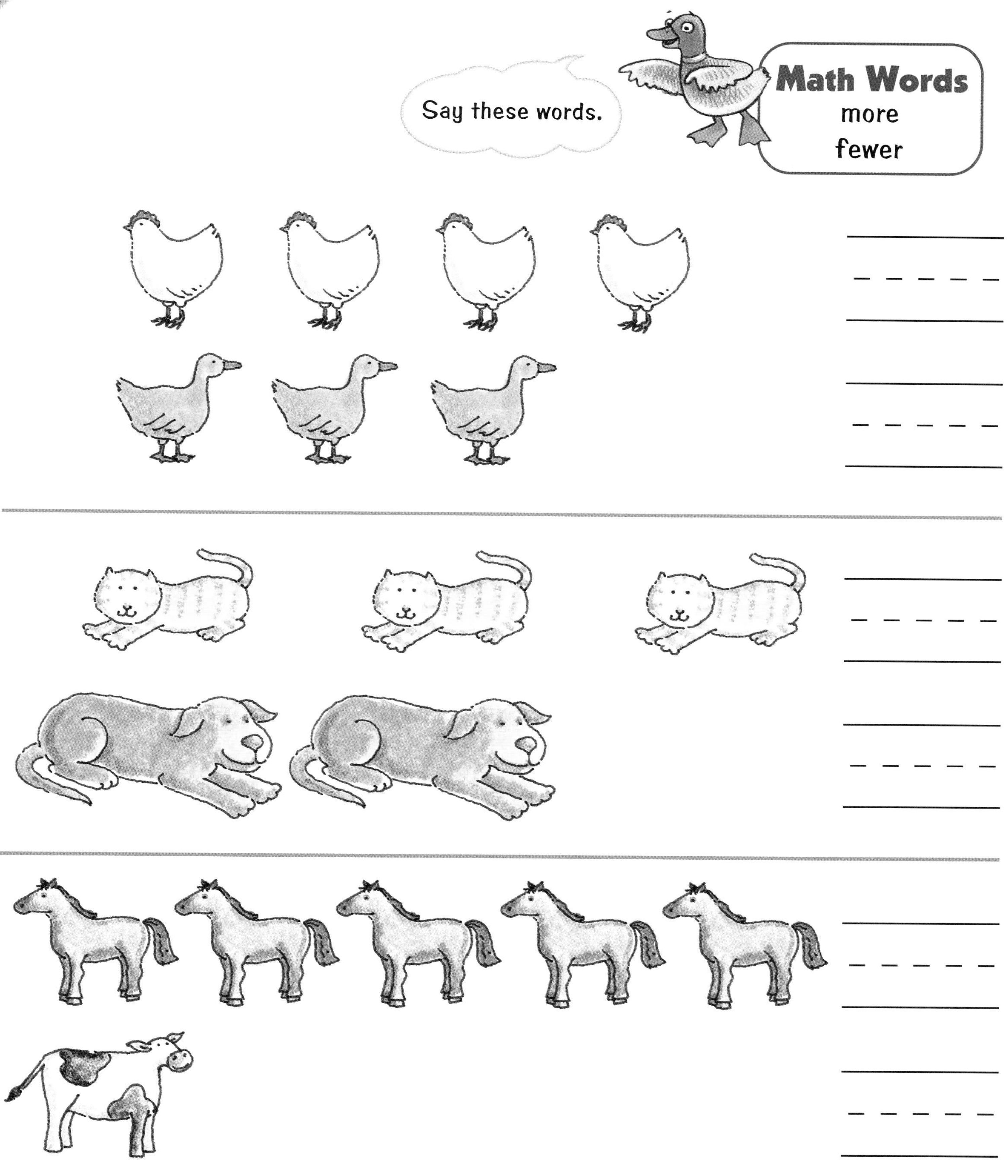

Draw lines to match the objects in each group. Write the number. Circle the group that has more.

Math at Home: Your child compared groups of objects by counting, circling and using the word more.
Activity: Have your child explain how he or she knew which group to draw a circle around each time.

Draw lines to match the objects in each group. Write the number. Circle the group that has fewer.

Name ___________________________________

____

____

Color a picture on the graph to show each object. Count the objects. Write how many.

Math at Home: Your child counted objects and made a picture graph with the information.
Activity: Ask your child how the graph tells us if there are more chickens or eggs.

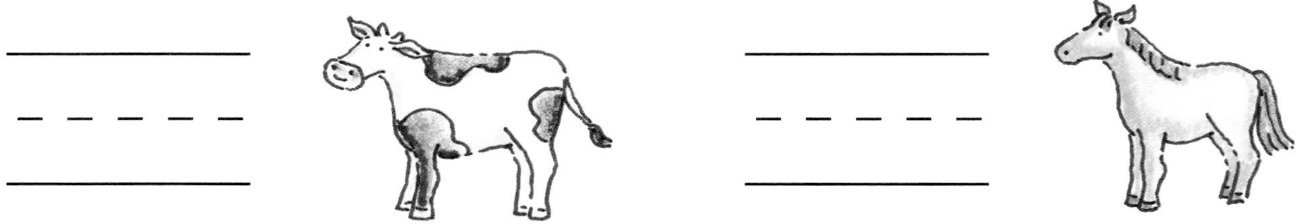

Look at the picture. Color a picture on the graph to show each animal.

Look at the graph. Count the pictures. Write each number. Are there more cows or horses? Circle the answer.

Name ______________________

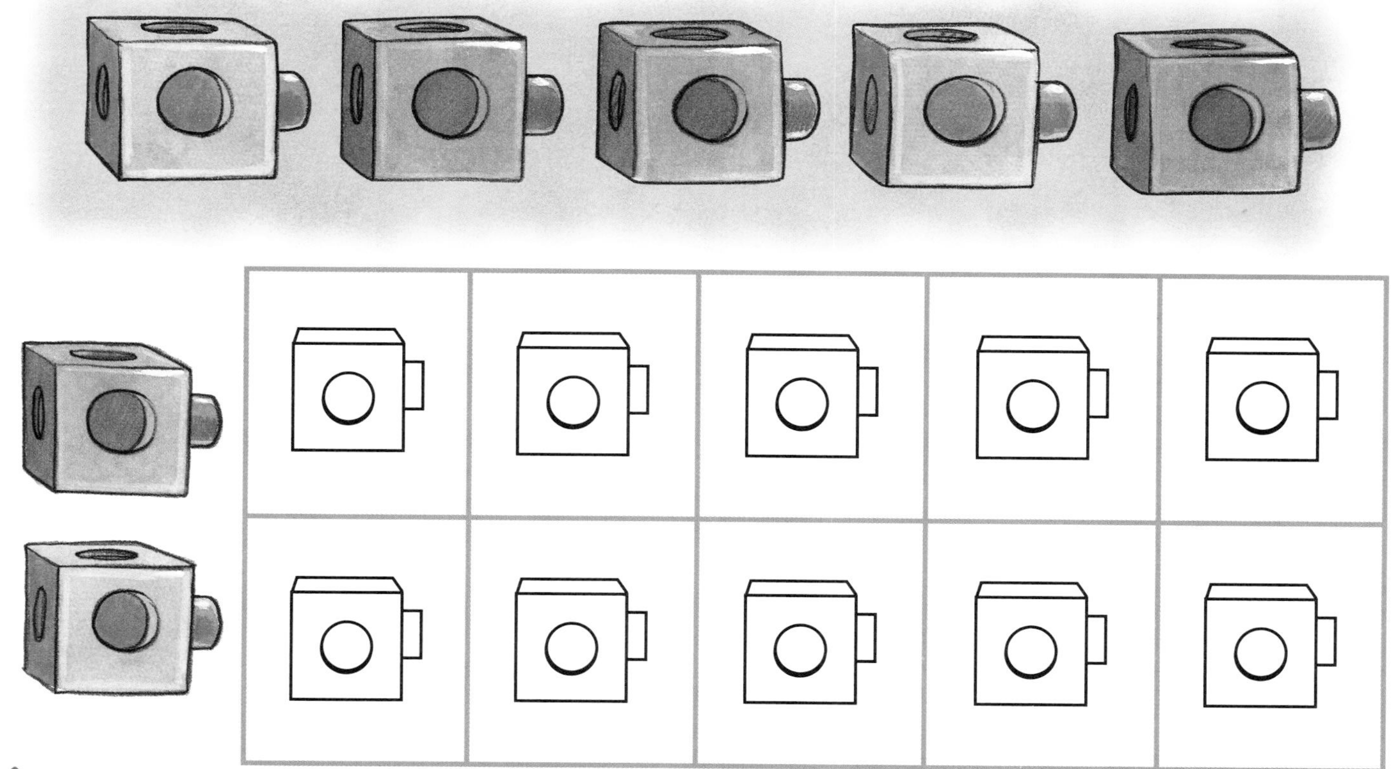

Count the animals. Write the number.

Color a picture on the graph to show each cube.

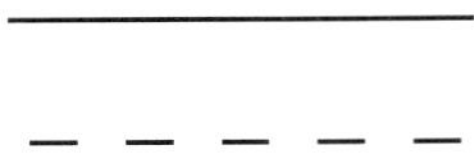

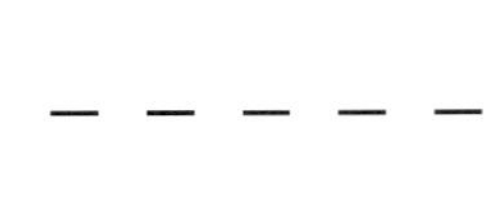

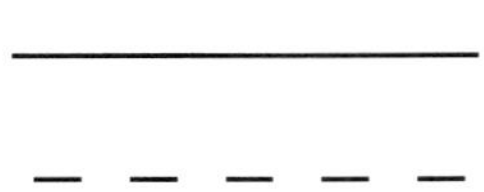

- Count how many. Write the number. Draw a group that shows more. Write the number.
- Count how many. Write the number. Draw a group that shows fewer. Write the number.
- Count how many. Write the number. Draw a group that is equal to the first group. Write the number.

CHAPTER

6 Numbers to 10

theme
Hops, Skips and Jumps

Use the Data

How many different kinds of animals do you see?

What You Will Learn

In this chapter you will learn how to:

- Count, read, and write numbers to 10.
- Compare and order numbers to 10.
- Use ordinal numbers to tenth.
- Make a table to solve problems.

Dear Family,
In Chapter 6, I will learn about the numbers 1-10. Here are some new vocabulary words and an activity that we can do together.

Counting to Ten

- On a sheet of paper, write the numbers 1–10 in a long row.
- Have your child make piles of 1–10 beans and place them on the paper above the correct number.

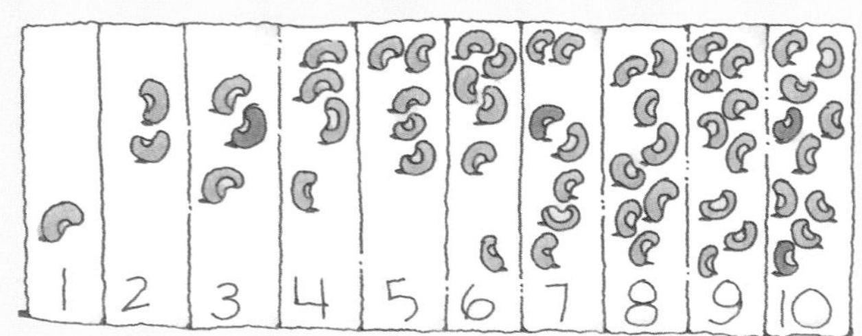

- Repeat the activity, but this time have your child count the beans from piles that you have made and then write the correct number.

use

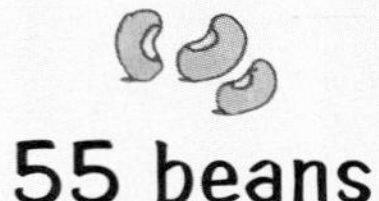

55 beans

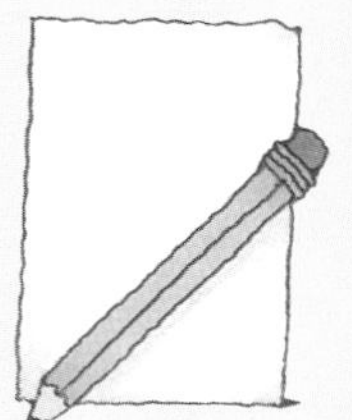

paper and pencil

Math Words

order

0, 1, 2, 3, 4, 5

These number are in order.

before

4 5

just before 5

between

5 6 7

between 5 and 7

after

6 7

just after 6

Name ____________________

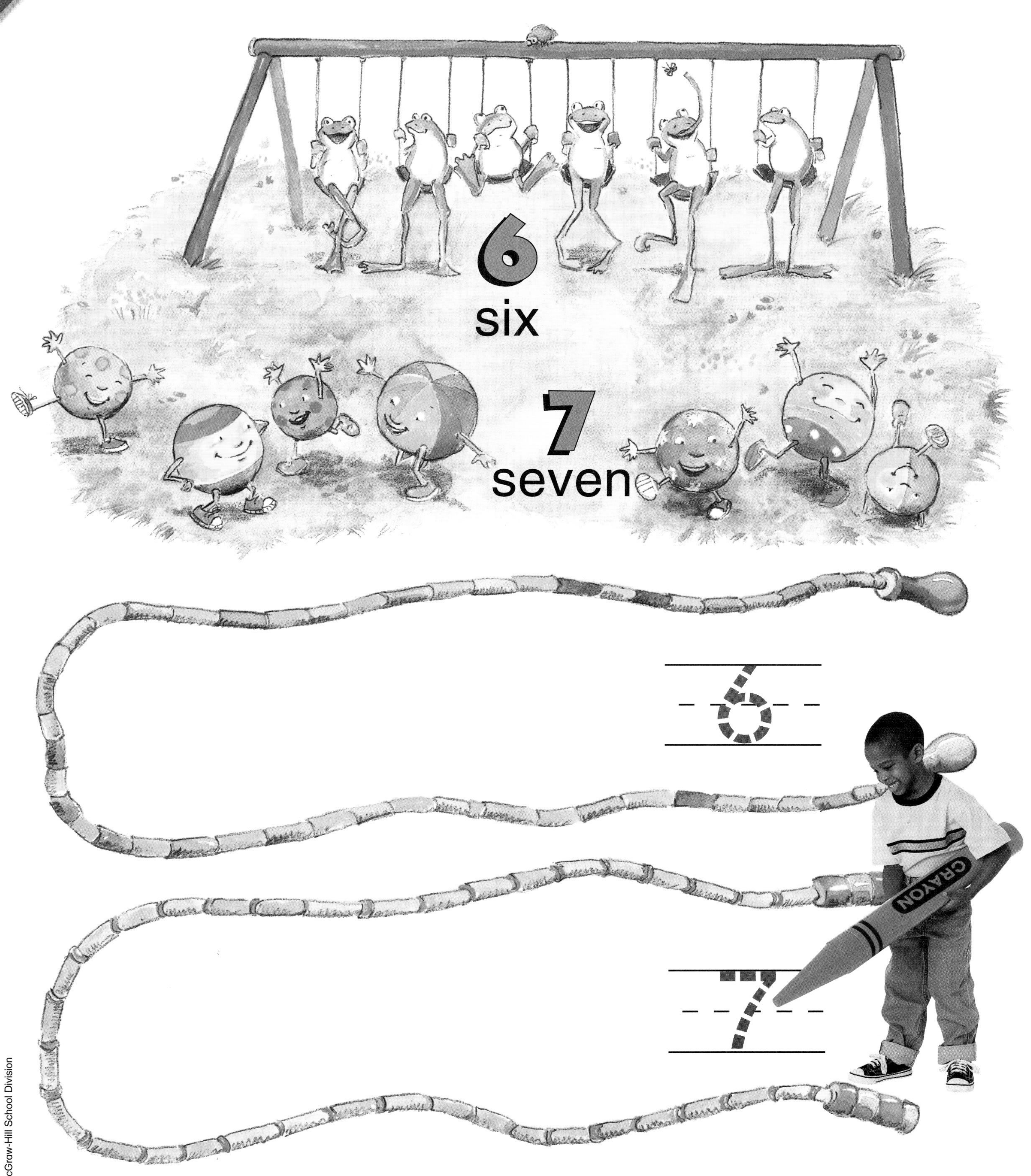

Use cubes to show 6 and 7. Draw the cubes. Write the number.

Math at Home: Your child is learning to count, read and write the numbers 6 and 7.
Activity: Have your child show you a group of 6 beans and a group of 7 beans.

Count the objects. Write the number.

Name ________________________________

8
eight

9
nine

8

9

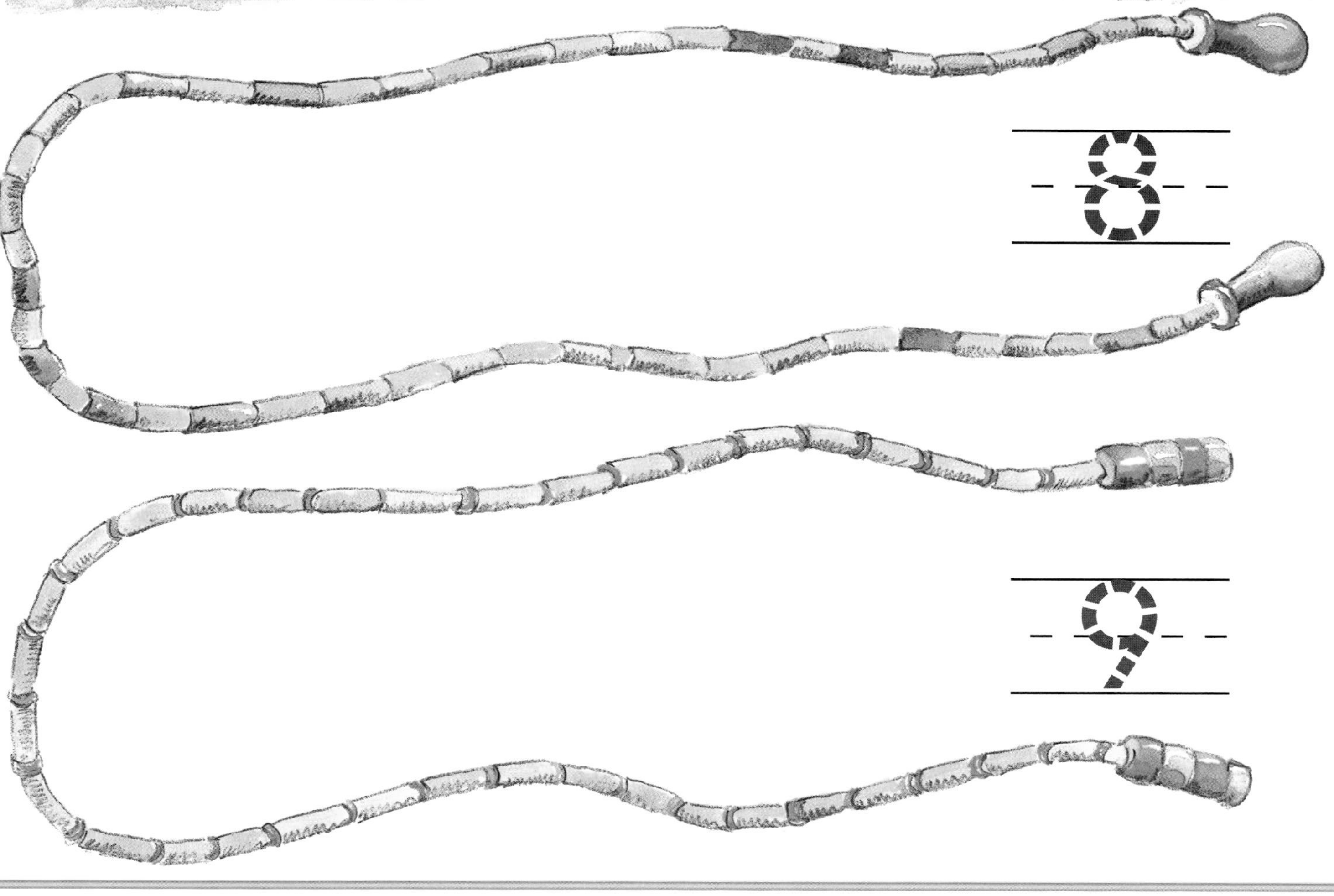

Use cubes to show 8 and 9. Draw the cubes. Write the number.

Math at Home: Your child is learning to count, read and write the numbers 8 and 9.
Activity: Have your child make a group of 8 buttons and then a group of 9 buttons.

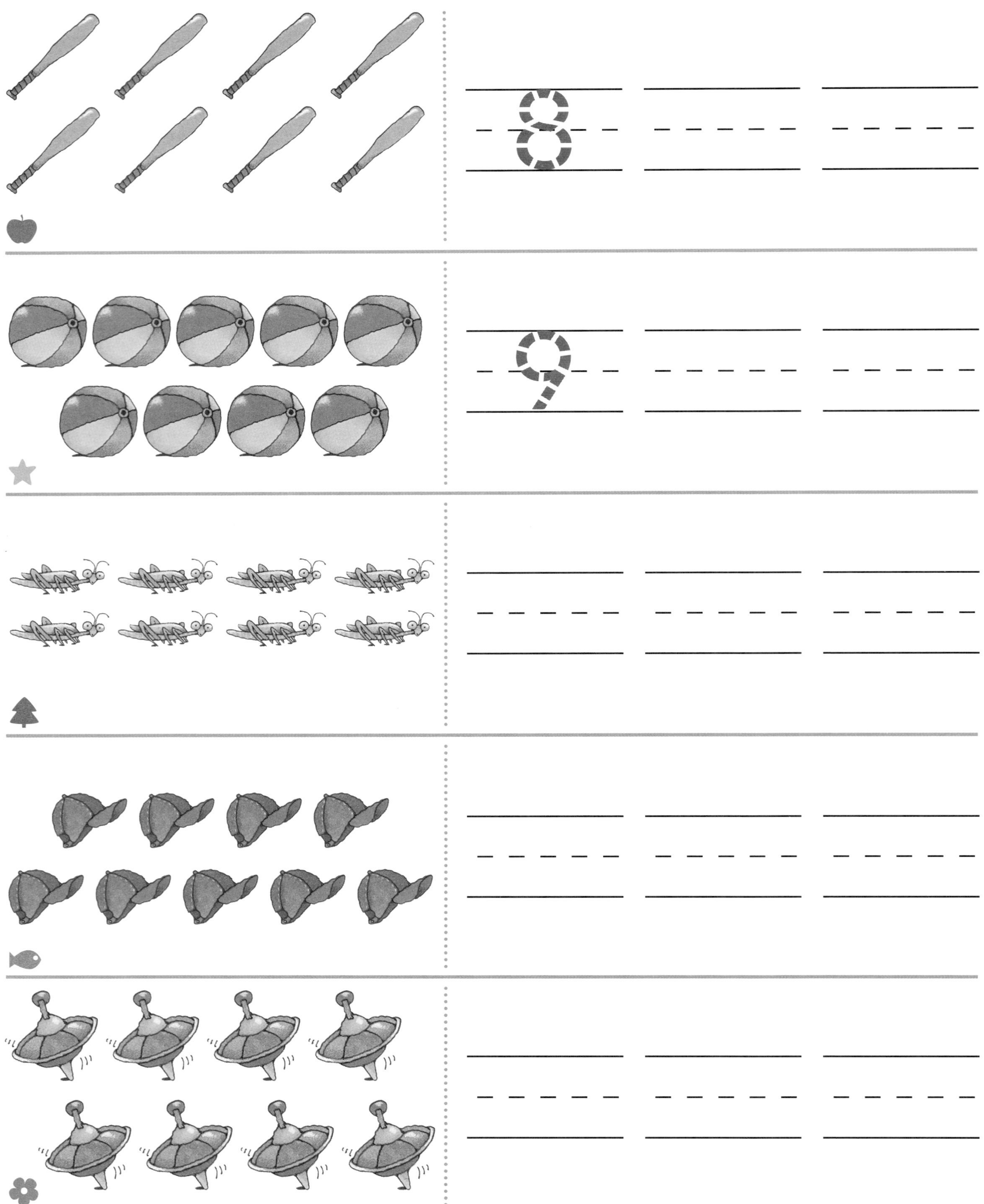

Count the objects. Write the number.

Name ______________________________

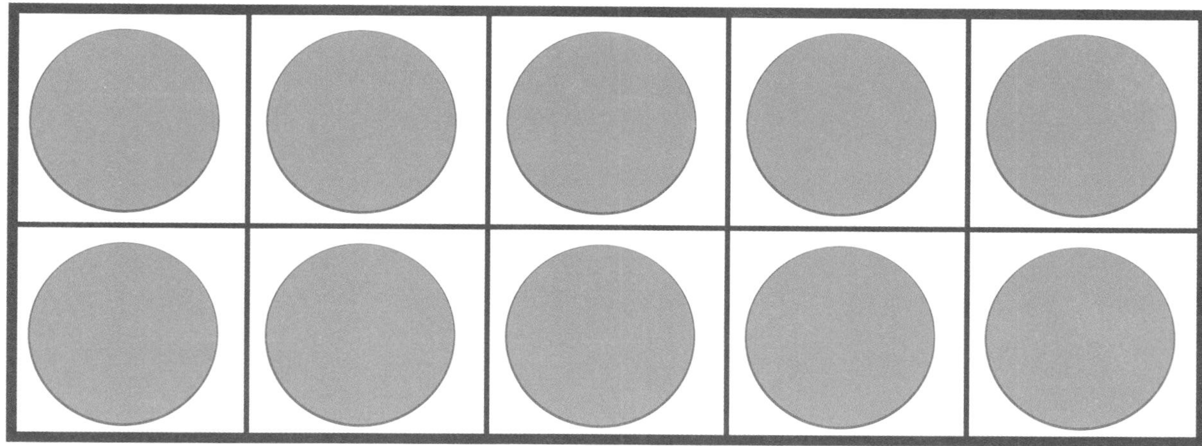

10

ten

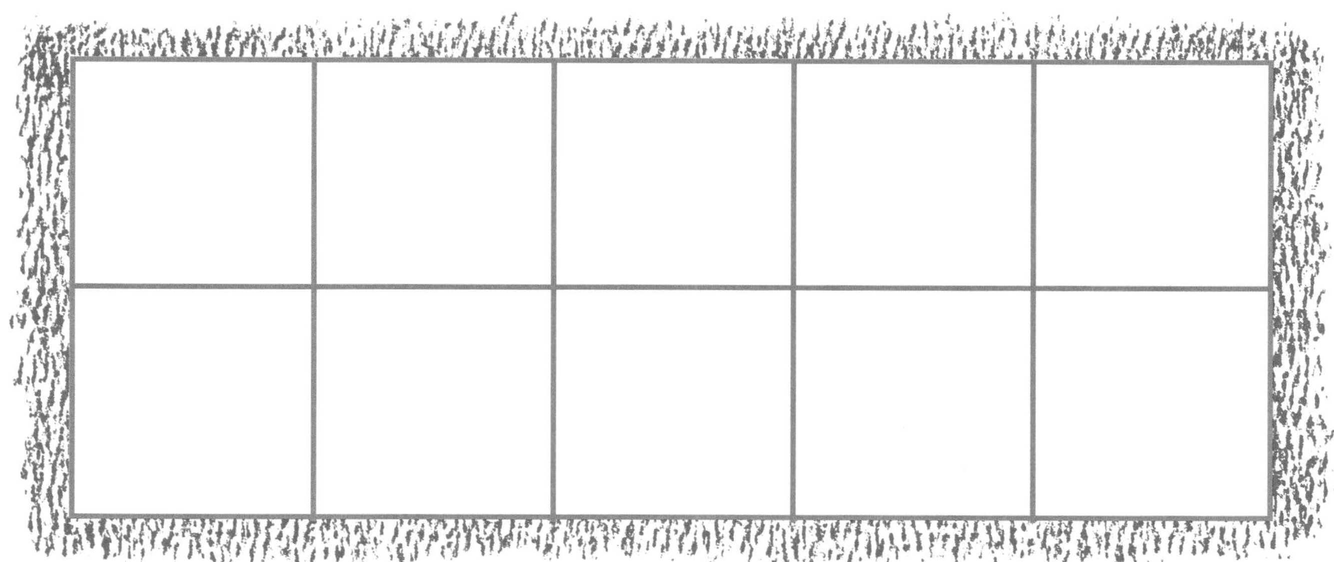

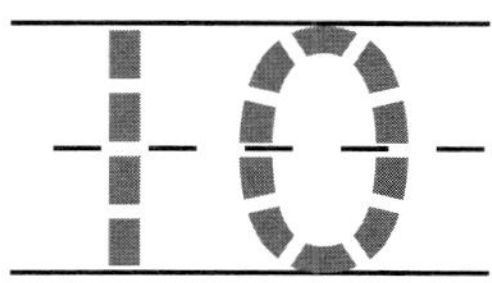

Place counters in each space on the ten-frame. Count how many. Draw the counters. Write the number.

Math at Home: Your child is learning to count, read, and write the number 10.
Activity: Ask your child to count 10 items such as paper clips or pennies.

Count the objects. Write the number.

Name____

Number Practice

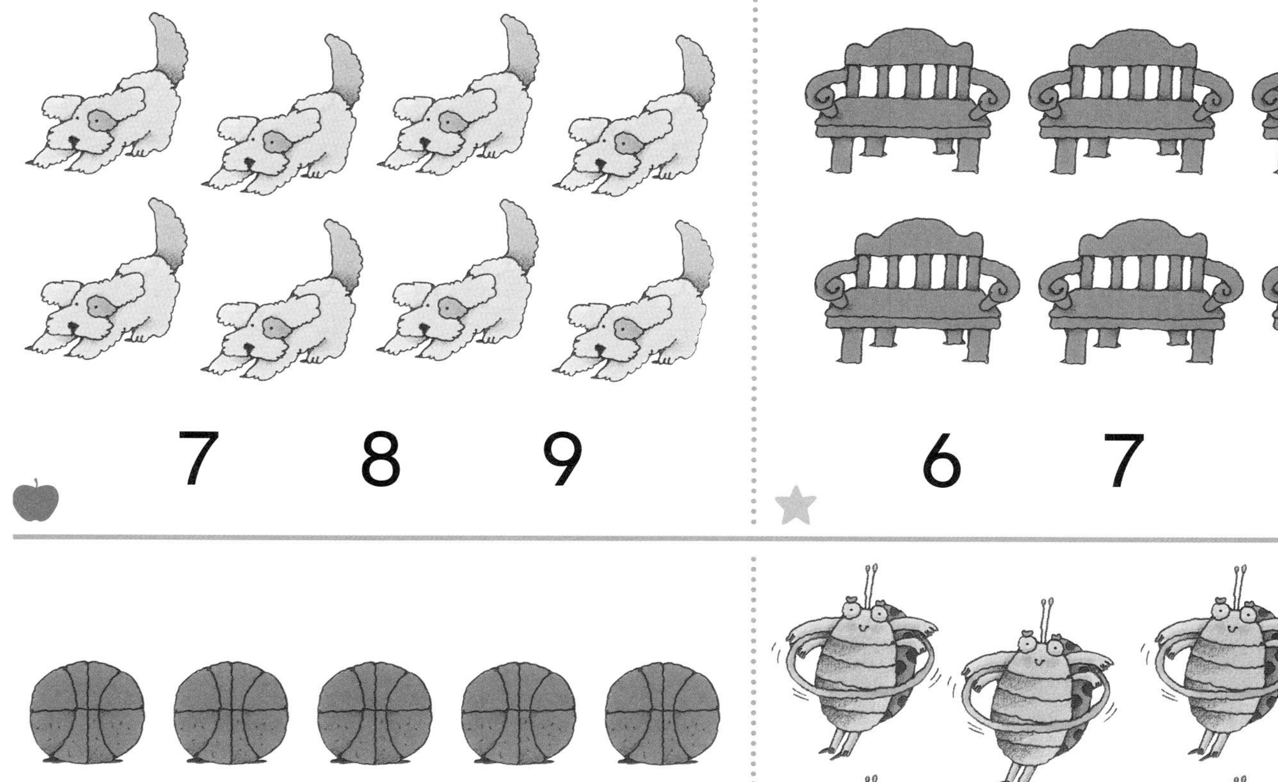

7 8 9

6 7 8

8 9 10

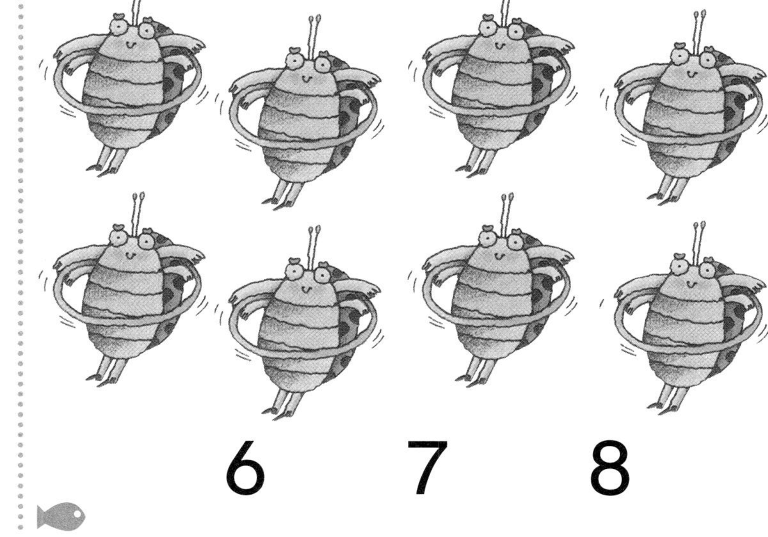

6 7 8

7 8 9

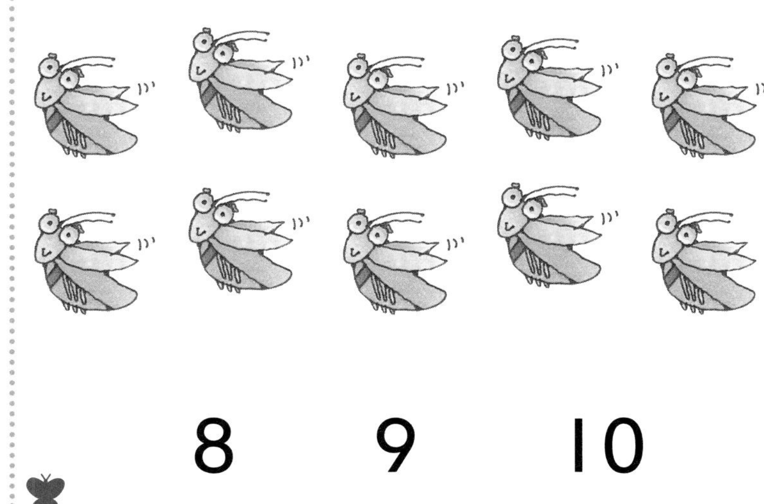

8 9 10

Count the objects. Circle the number.

Math at Home: Your child practiced numbers to 10.
Activity: Make groups of small items such as pennies to show 6, 7, 8, 9, and 10. Ask your child to count each group and write the number.

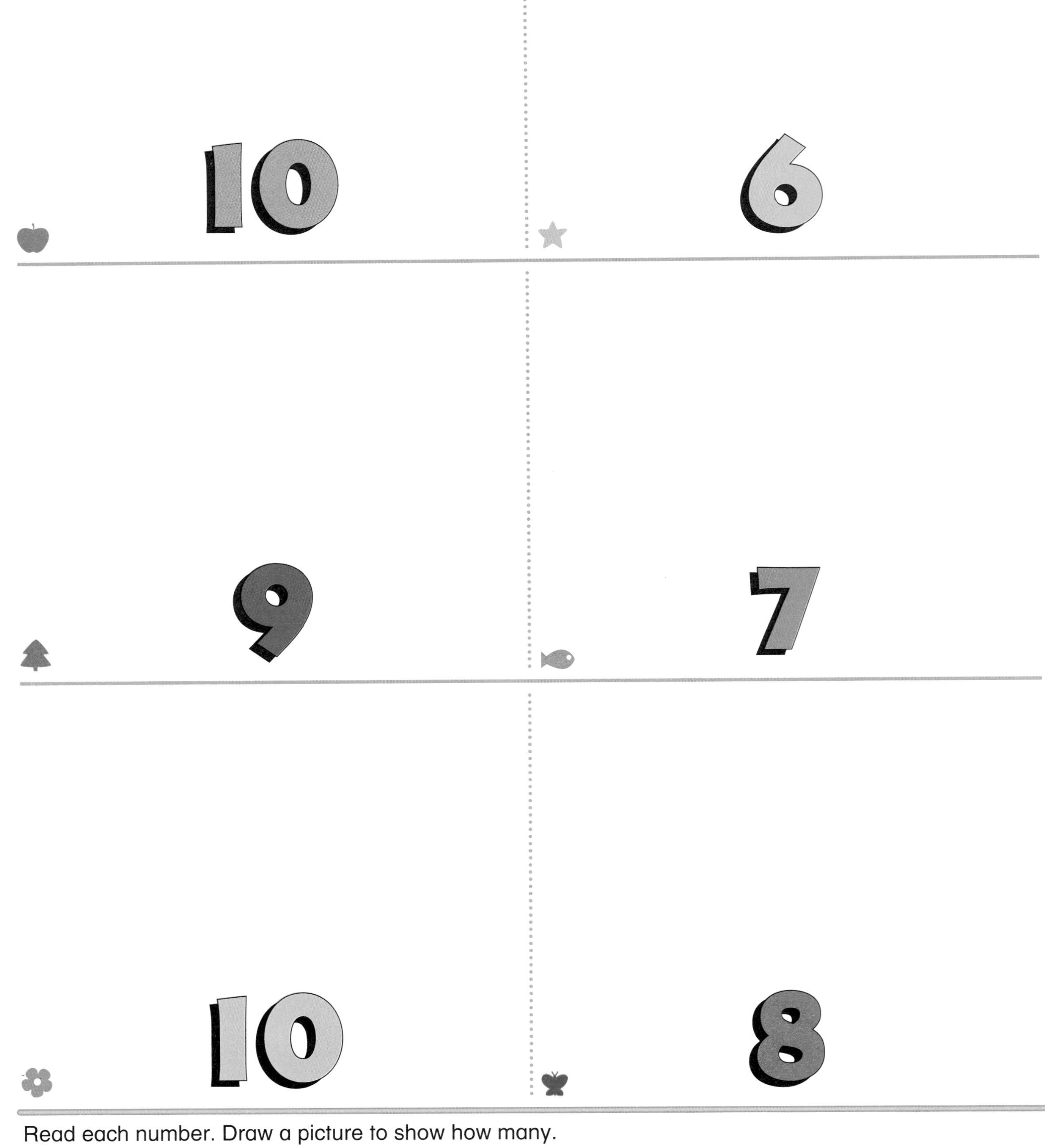

Read each number. Draw a picture to show how many.

Name______________________

Math Words
tally mark
table

Make a Table

Say these words.

Count the animals. Make a tally mark on the chart to show each animal. Write the numbers. Circle the number that is fewer.

Math at Home: Your child made a table to solve problems.
Activity: Have your child tell how he or she knew whether there are fewer kangaroos or kittens.

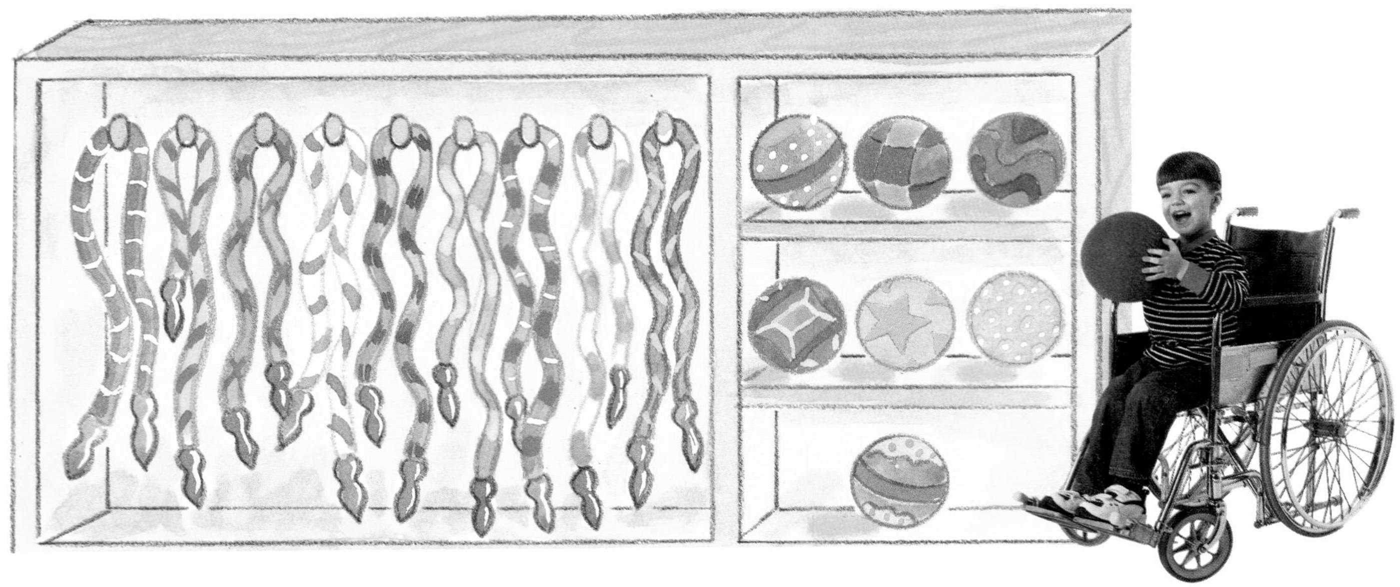

- Count the toys. Make a tally mark on the chart to show each toy.
- Write the numbers. Circle the number that is more.

Name ______________________

	one more	one fewer
	___	___
	___	___
	___	___

Look at the objects in the first box. Draw a picture to show one more and one fewer. Write the numbers.

Math at Home: Your child is learning to compare numbers to 10.
Activity: Show your child a group of 4 or 5 of the same items. Ask him or her to make a group with one more and a group with one fewer than the first group.

Draw lines to match the objects in each group. Write the number. Circle the number that is more.

Draw lines to match the objects in each group. Write the number. Circle the number that is fewer.

Name ______________________________

Math Words
order
before
between
after

1 2 3 4 5 6 7 8 9 10

Color cubes to show each number.

Math at Home: Your child is learning to order numbers from 0–10.
Activity: Ask your child to count aloud from 0–10 several times a day.

4 □ 6 7 8 □

1 2 □ 4 □ 6

□ 6 7 8 9 □

Write the missing numbers.

Name________________________________

Color a box on the graph to show each animal. Count the animals. Write how many.

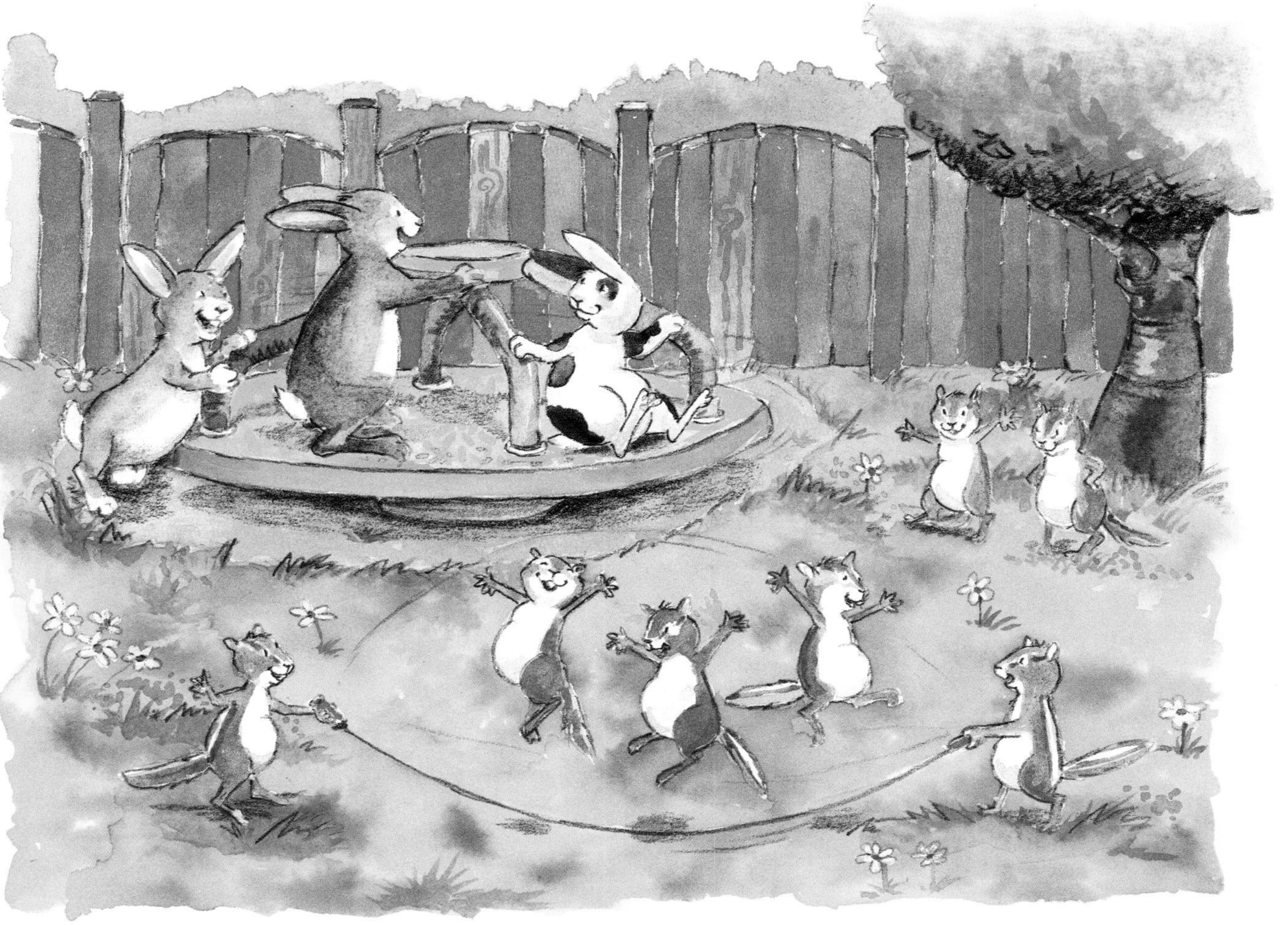

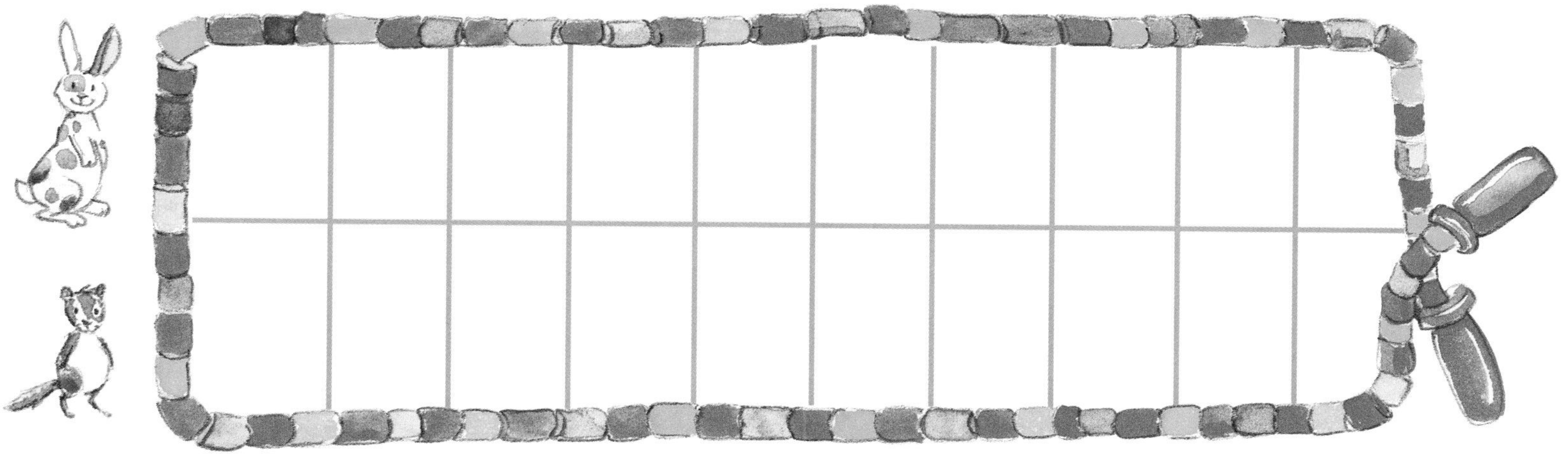

Math at Home: Your child counted animals and made a bar graph with the information.
Activity: Ask your child how the graph tells us if there are more bunnies or chipmunks.

- Color a box on the graph to show each animal.
- Count the animals. Write how many. Are there more birds or ladybugs? Circle the answer.

Name ______________________________

Ordinal Numbers to 10

Say these words.

Math Words
first
next
last

- Color the first inner tube red. Color the next inner tube blue. Color the last inner tube green.
- Draw a circle around the second child. Draw a box around the fourth child. Draw a line under the fifth child.

Math at Home: Your child is learning ordinal numbers to 10.
Activity: Place 5 of your child's toys in a row on the bed. Ask your child to identify the first, second, and last toy.

Draw a circle around the seventh kangaroo. Draw a circle around the third ant.
Draw a circle around the ninth frog. Draw a circle around the sixth duck.
Draw a circle around the tenth bunny.

Name ______________________

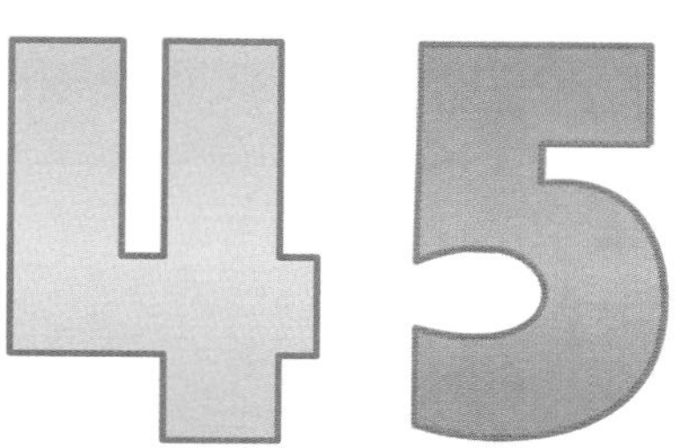

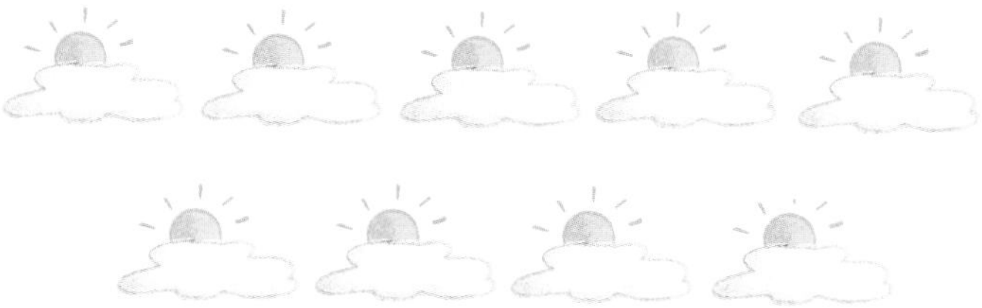

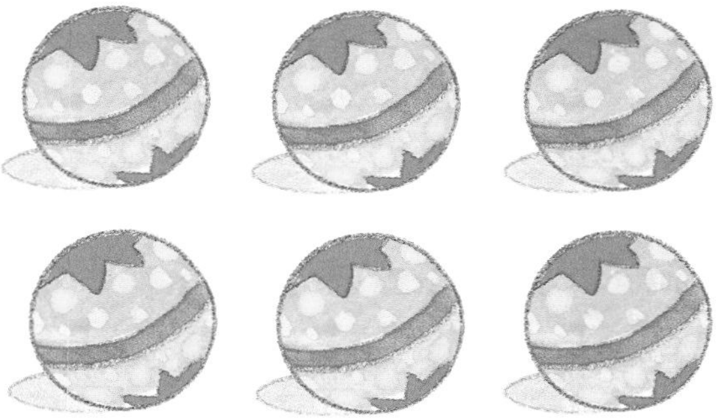

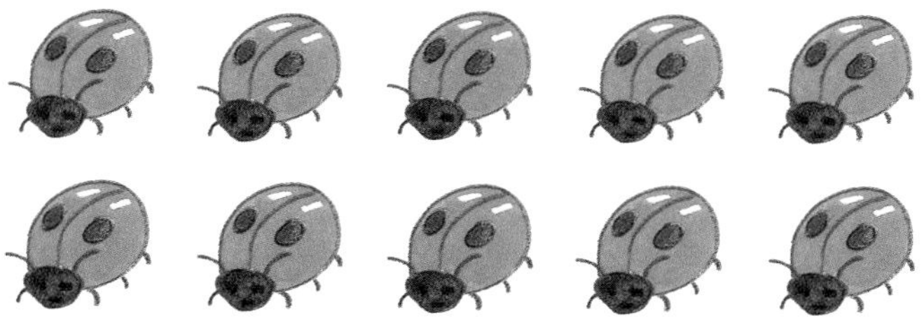

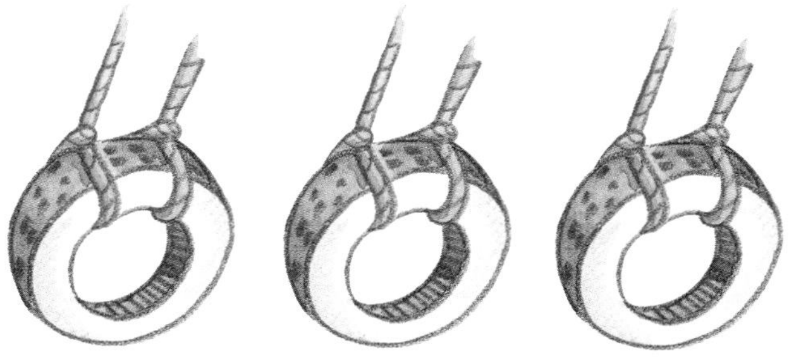

Count how many. Write the number.

Write the missing numbers.

- Draw lines to match the objects in each group. Write the number. Circle the number that is more.
- Color the second dog green. Color the fifth dog red.
- Color a box on the graph to show each ball. Count the boxes. Write the numbers.

CHAPTER 7

Numbers to 20

theme

Around the Neighborhood

Use the Data

How many spots do you see on the puppies?

What You Will Learn

In this chapter you will learn how to:

- Count, read, and write numbers to 20.
- Compare and order numbers to 20.
- Skip count by twos to 20.
- Draw a picture to solve problems.

Dear Family,
In Chapter 7, I will learn about numbers 11 to 20. Here are some new vocabulary words and an activity that we can do together.

Toothpicks to Twenty

Math Words

number

3
A number tells how many.

count

1, 2, 3
Count to find how many.

skip count

2, 4, 6, 8, 10

- Write the even numbers from 2 to 20 on a sheet of paper. For example: 2, 4, 6, 8.....

2 4 6 8 10 12 14 16 18 20

- Have your child make groups of two toothpicks and place them above each number.
- Count by twos with your child while pointing to each group of toothpicks.

use

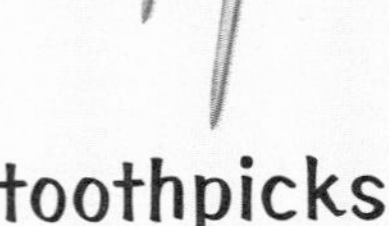

toothpicks

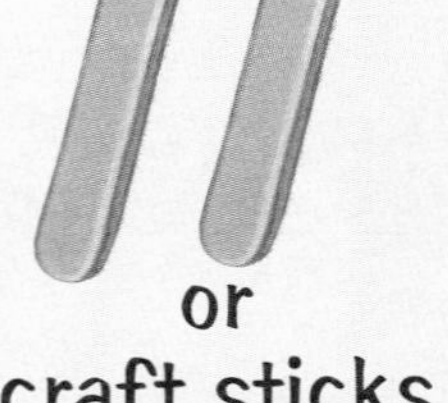

or
craft sticks

paper and pencil

Additional activities at www.mhschool.com/math

Name________________________

Math Words
tens
ones

Say these words.

eleven

11

twelve

12

thirteen

13

Use counters and the 10-frame to show each number. Draw the counters. Write each number.

Math at Home: Your child made groups of 11, 12, and 13.
Activity: Show your child up to 13 paper clips or beans. Have him or her arrange the objects as groups of tens and ones.

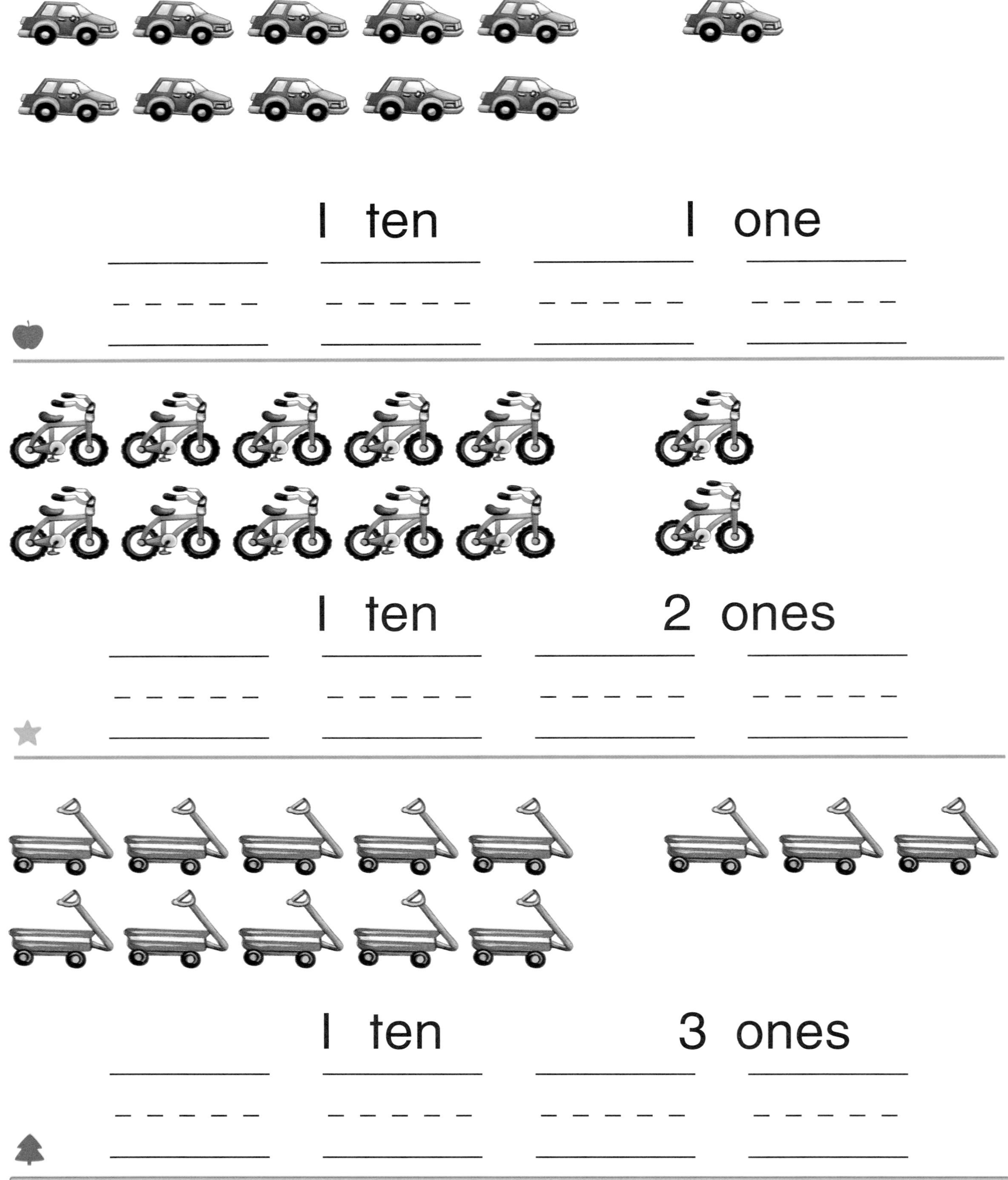

Count the objects. Write the number.

Name ______________________

fourteen
1 ten 4 ones

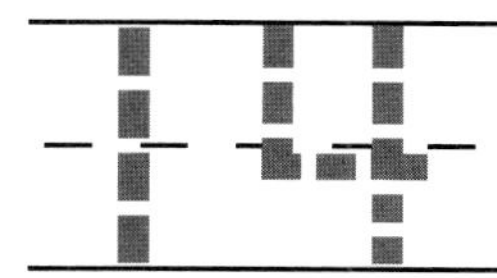

fifteen
1 ten 5 ones

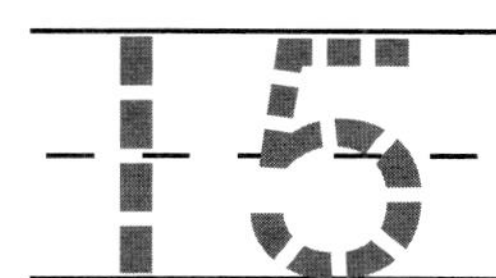

Use counters and the 10-frame to show tens and ones.
Draw the counters. Write each number.

Math at Home: Your child made groups of 14 and 15.
Activity: Show your child up to 15 paper clips or beans. Have him or her arrange the objects as groups of tens and ones.

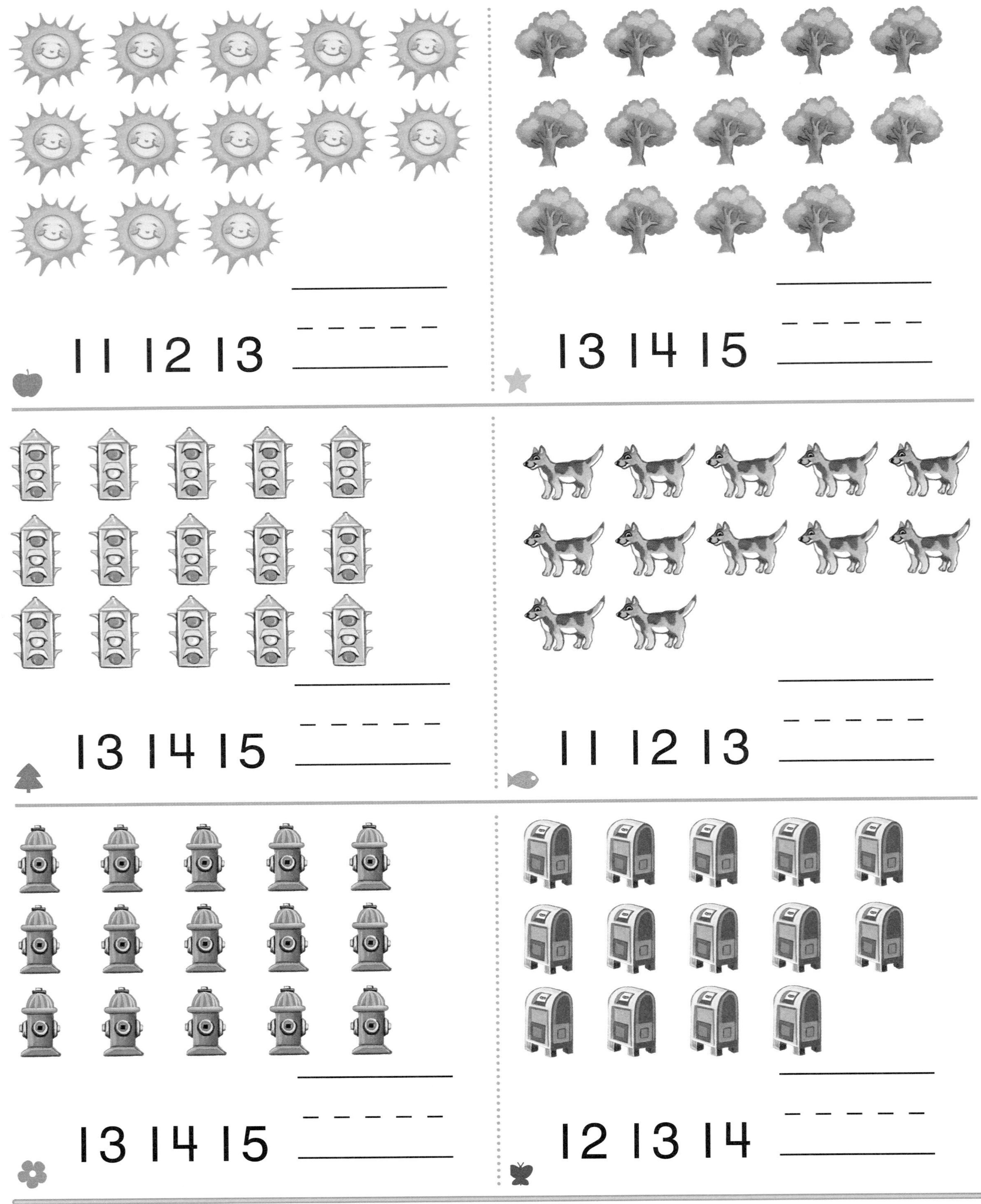

Count the objects. Circle the number. Write the number.

Name____

sixteen
1 ten 6 ones

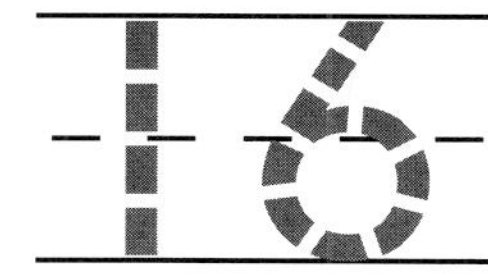

seventeen
1 ten 7 ones

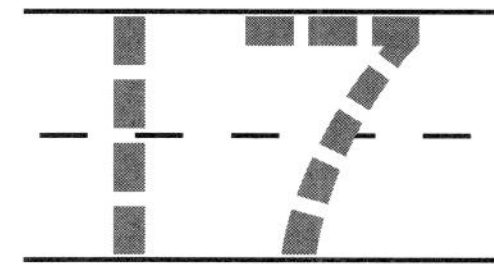

Use counters and the 10-frame to show tens and ones. Draw the counters. Write each number.

Math at Home: Your child made groups of 16 and 17.
Activity: Show up to 17 paper clips or beans. Have your child arrange the objects as groups of tens and ones.

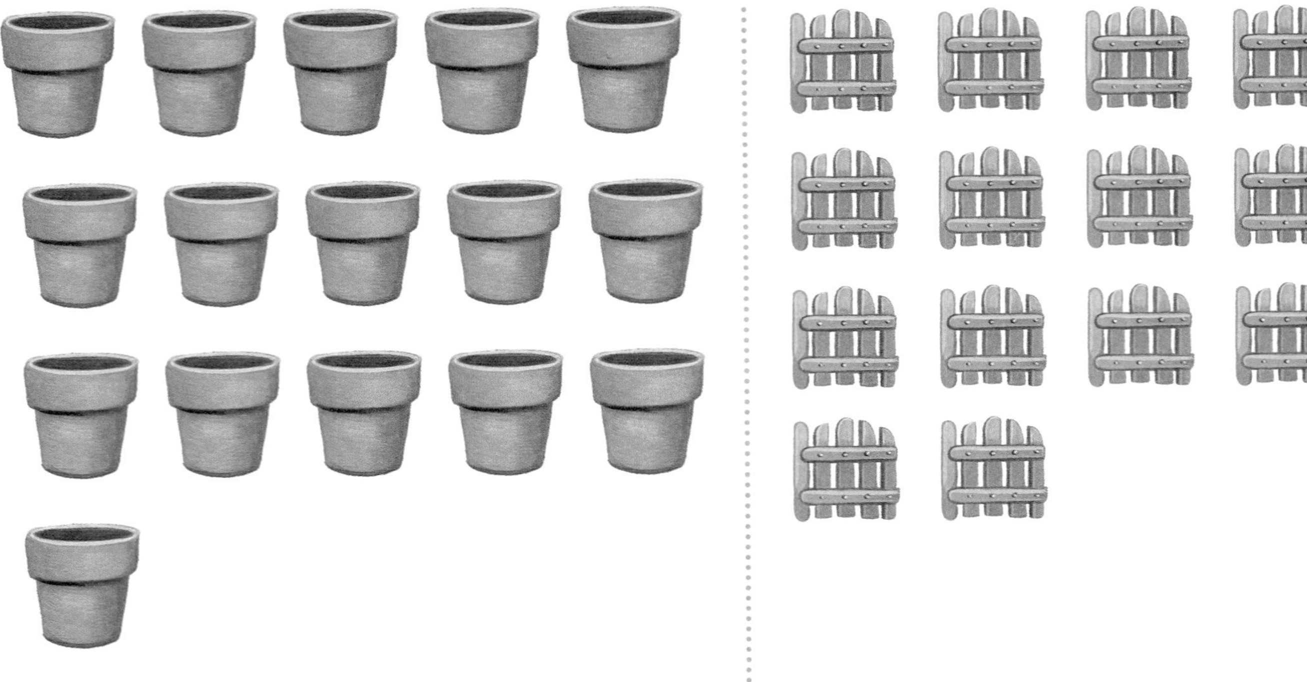

15 16 17

15 16 17

10 11 12

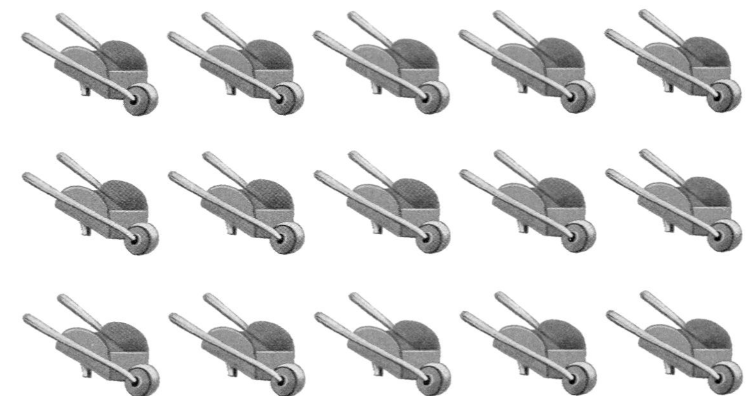

13 14 15

Count the objects. Circle the number. Write the number.

Name______________________________

eighteen
1 ten 8 ones

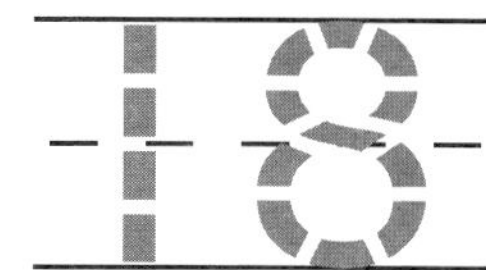

nineteen
1 ten 9 ones

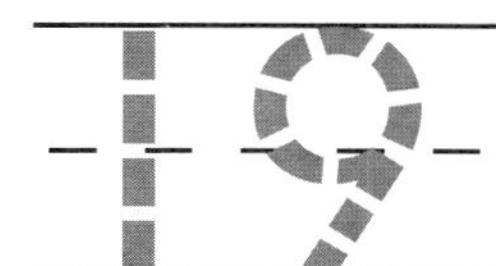

Use counters and the 10-frame to show tens and ones. Draw the counters. Write each number.

Math at Home: Your child made groups of 18 and 19.
Activity: Show up to 19 buttons or pebbles. Have your child arrange the objects as groups of tens and ones.

17 18 19

16 17 18

15 16 17

11 12 13

Count the objects. Circle the number. Write the number.

Name____

twenty
2 tens

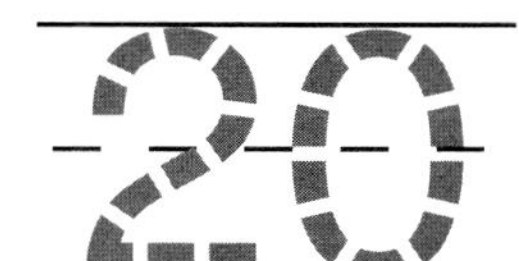

Use counters and the 10-frame to show 2 tens. Draw the counters. Write the number.

Math at Home: Your child made groups of 20.
Activity: Show up to 20 buttons or pebbles. Have your child arrange the objects as groups of tens and ones.

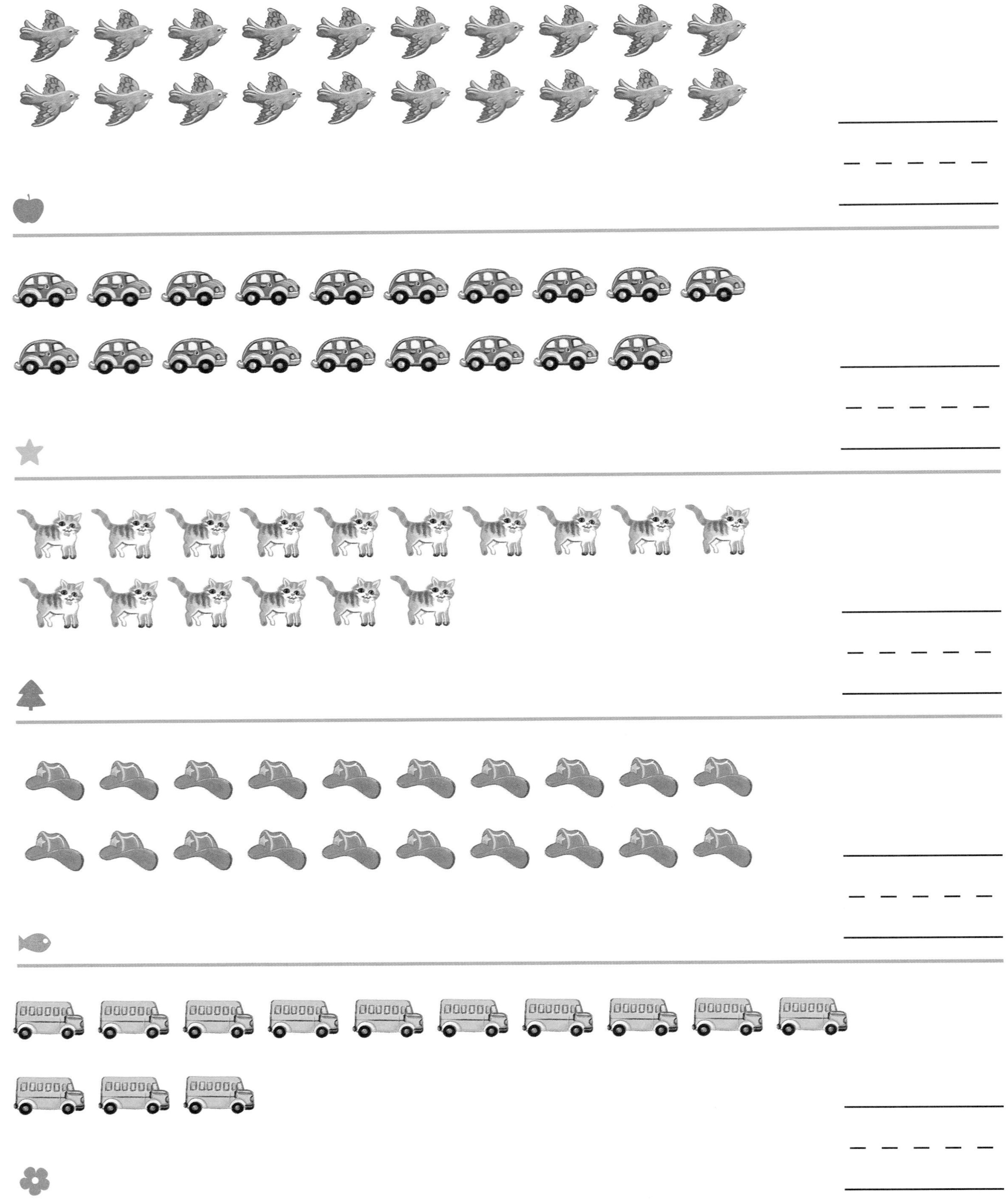

Count the objects. Write the number.

Name ______________________

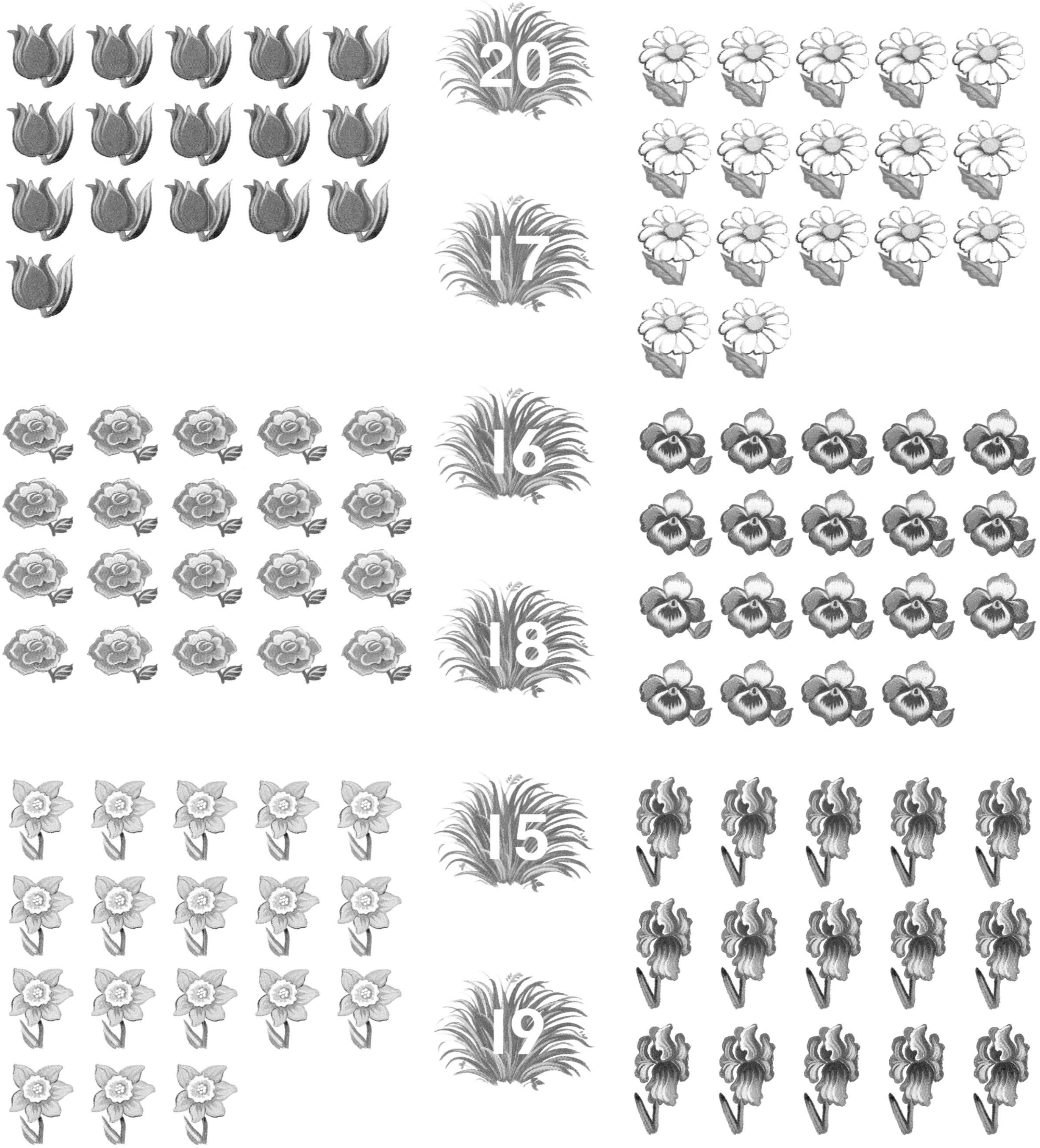

20

17

16

18

15

19

Draw lines to match groups and numbers.

Math at Home: Your child is practicing numbers 11-20.
Activity: Place buttons or other small items in groups of 11-20. Have your child count the objects and write the numbers.

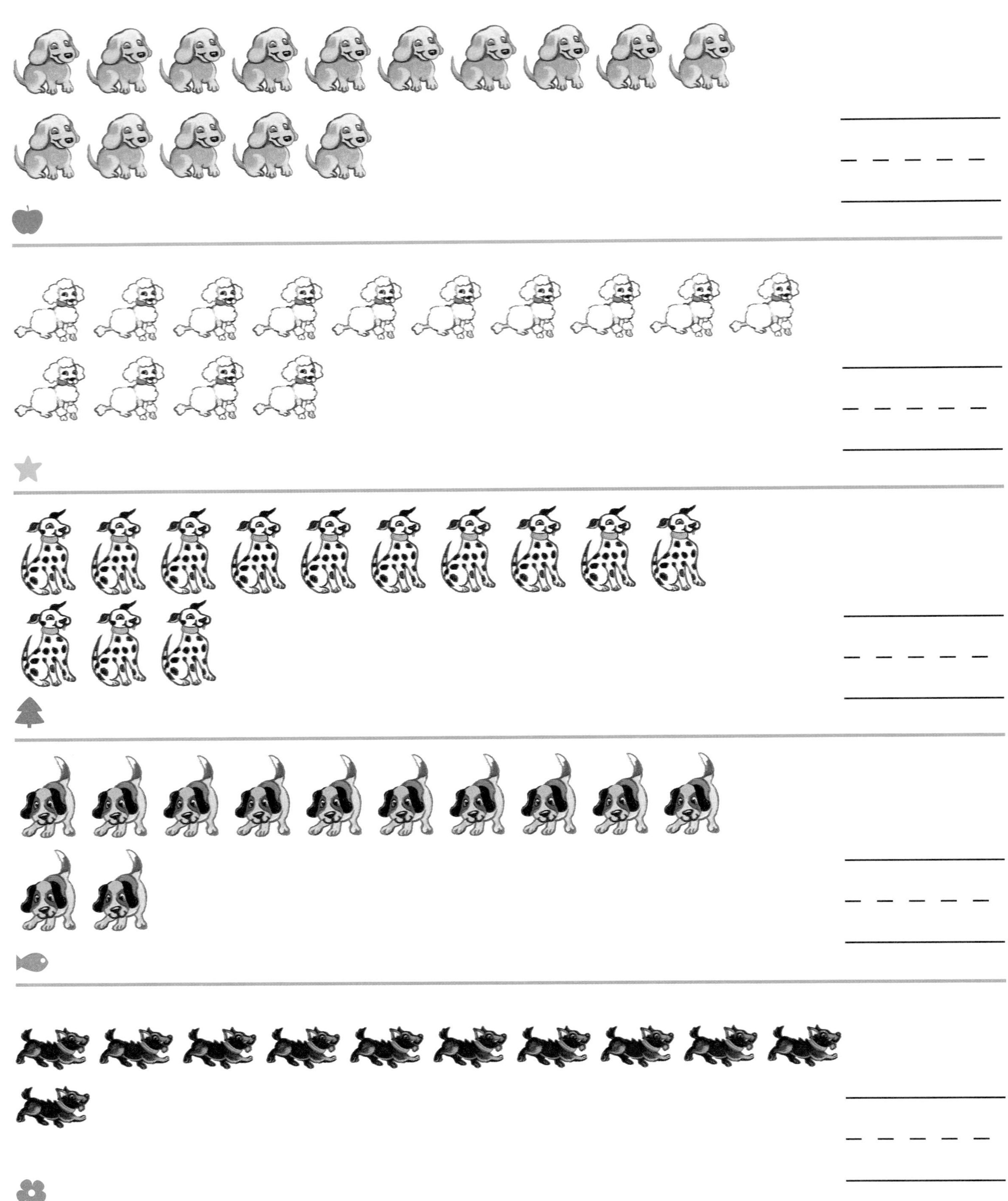

Count the objects. Write the number.

Name ____________________

Draw a Picture

Look at the picture. Draw a bug on each flower.

Math at Home: Your child drew pictures to solve a problem.
Activity: Ask your child to tell how he or she knew how many bugs to draw on the flowers.

Look at the picture. Draw a water dish for each puppy.

Compare Numbers to 20

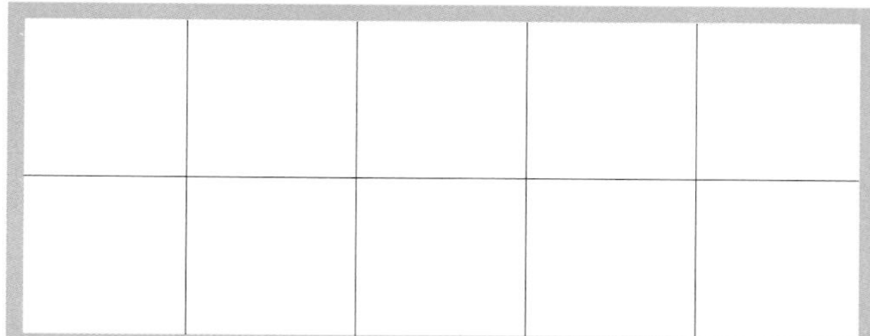

I ten 4 ones

I ten 6 ones

I ten 2 ones

I ten I one

Use the 10 frames. Draw circles to show each number. Write the number.
Circle the number that is more.

Math at Home: Your child compared numbers to 20.
Activity: Ask your child to tell how he or she knows which number is more.

Count the objects. Write the numbers. Circle the number that is fewer.

Name ______________________________

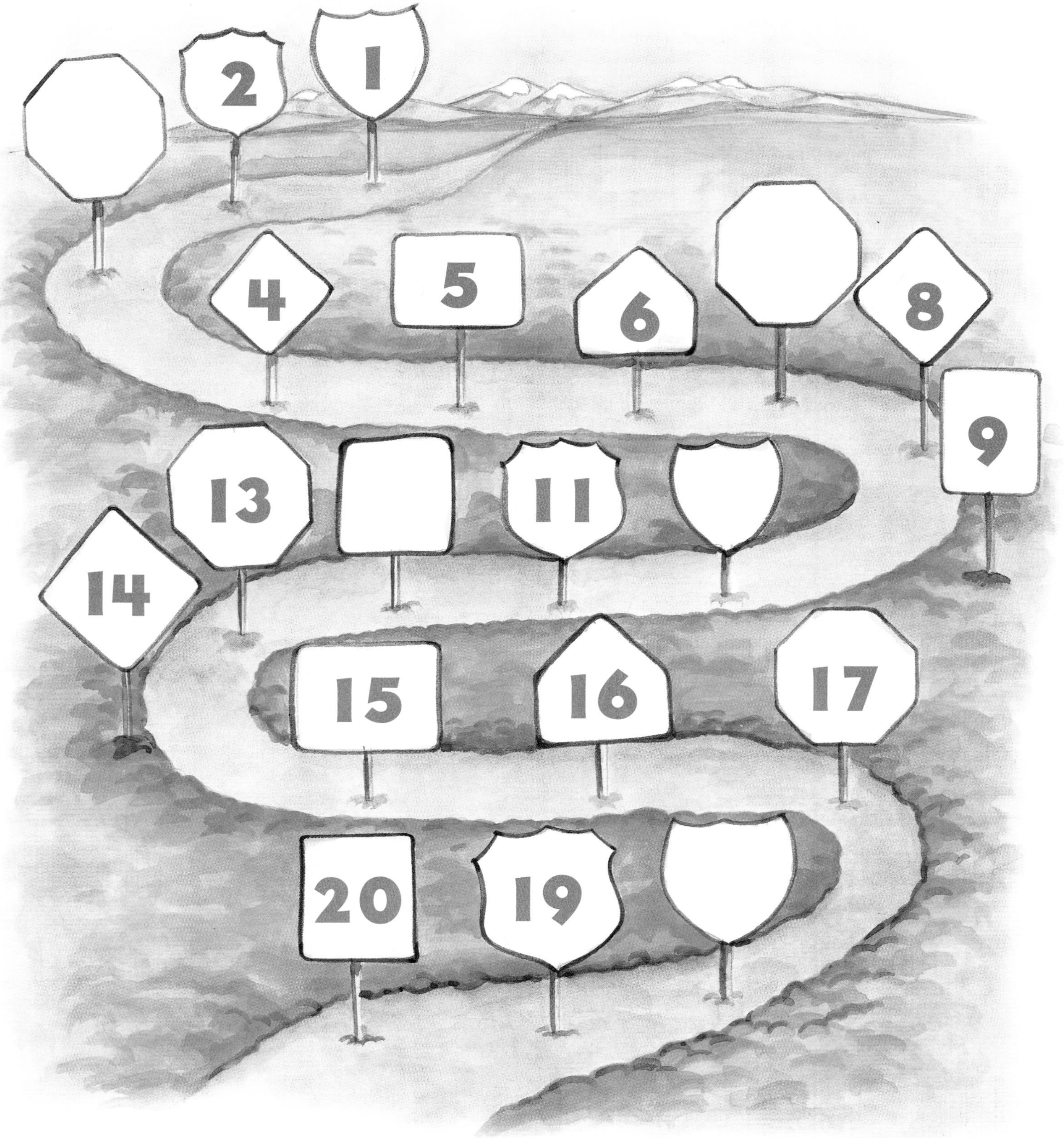

Write the missing numbers.

Math at Home: Your child is learning to order numbers to 20.
Activity: Point to a number on the page. Ask your child to tell the numbers before and after.

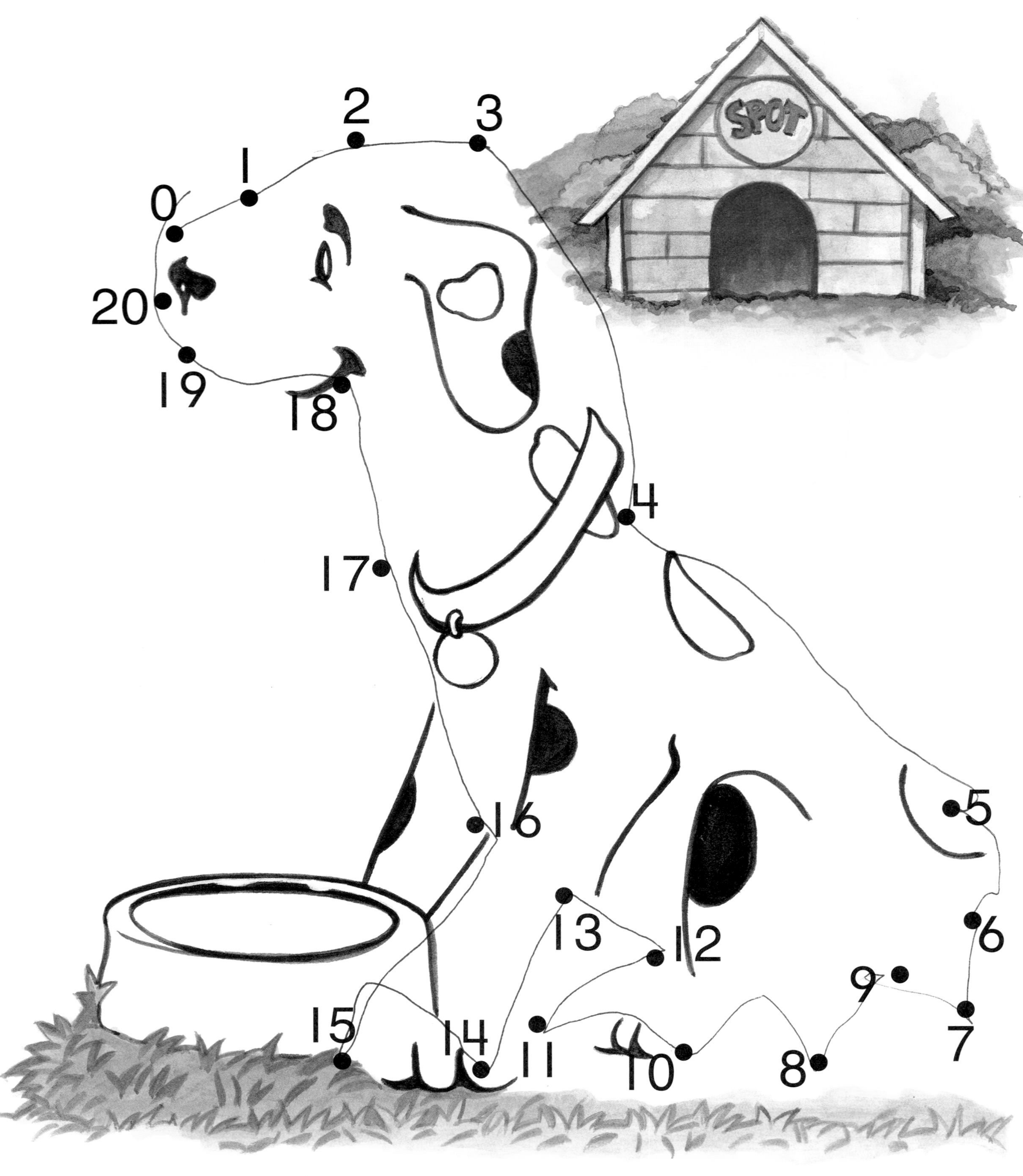

Connect the dots to show the numbers 0 to 20 in order. Color the picture.

Name ____________________

Math Word
skip count

0 1 2 3 4 5 6 7 8 9 10 11 12 13 14 15 16 17 18 19 20

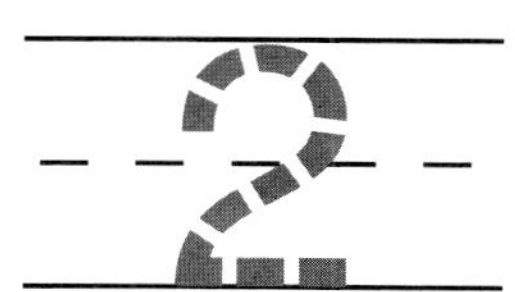

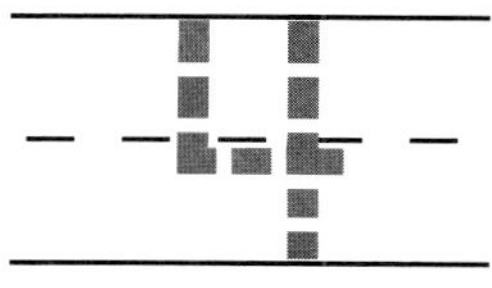

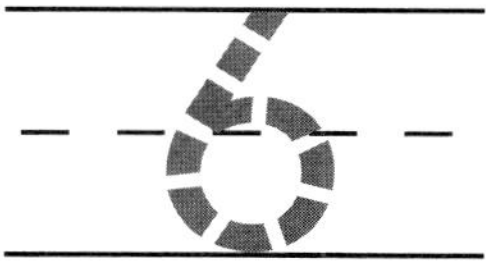

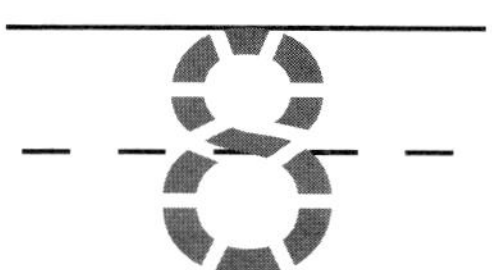

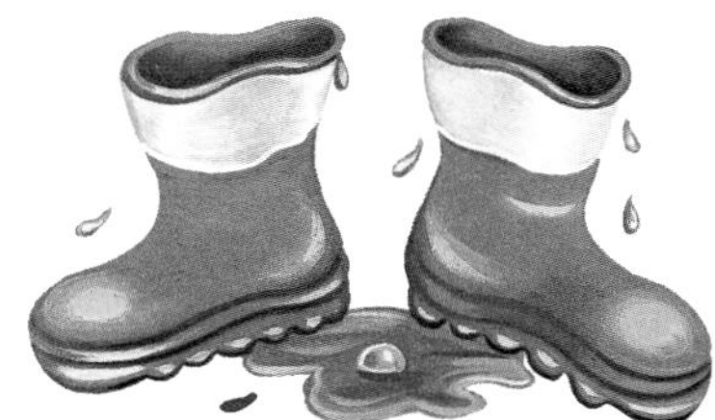

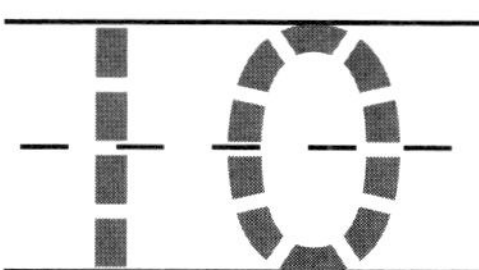

_____ in all

Skip count to count the boots. Write the numbers.

Math at Home: Your child is learning to skip-count by twos.
Activity: Line up pairs of shoes on the floor and have your child skip count to tell how many there are.

Connect the dots to skip count.

Name ______________________________

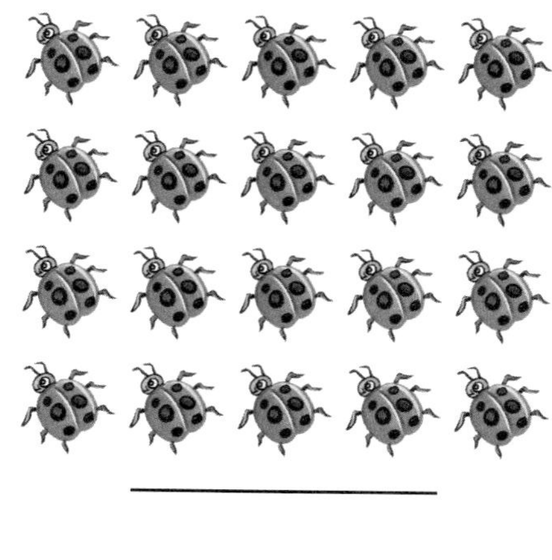

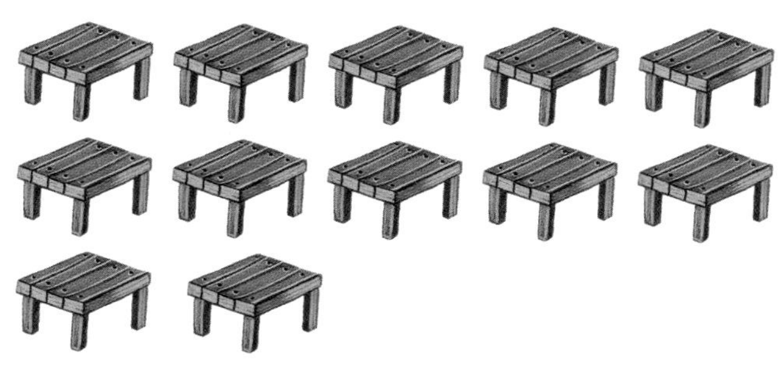

I ten 6 ones

Count the objects. Write the number.

Use the 10-frame. Draw circles to show the number. Write the number.

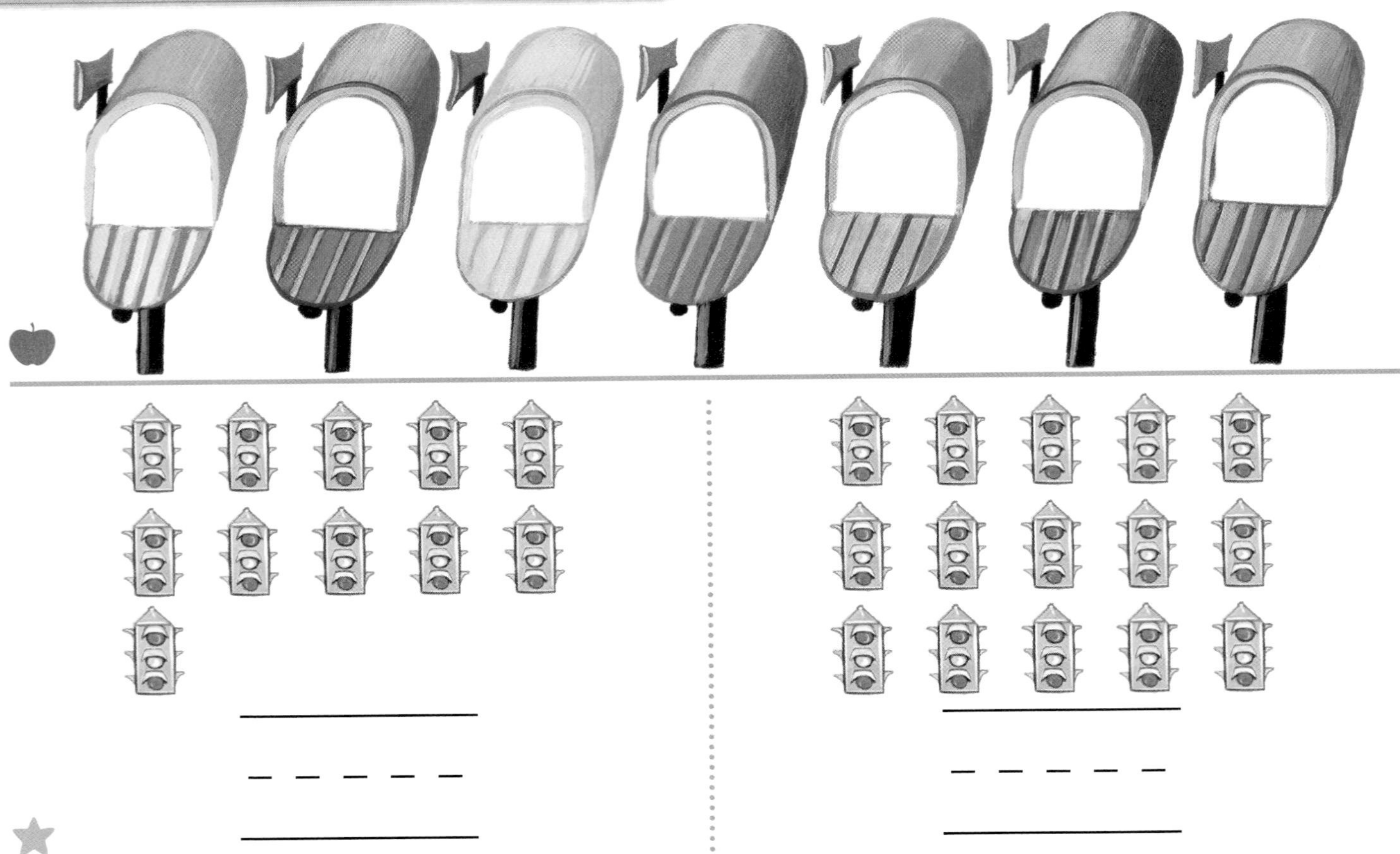

0 1 2 3 4 5 6 7 8 9 10 11 12 13 14 15 16 17 18 19 20

8, ☐, 10, 11, ☐, 13

Look at the picture. Draw a letter for each mailbox. Count the objects. Write the numbers. Circle the number that is more. Write the missing numbers. Skip count. Write the numbers.

CHAPTER 8

Numbers to 100

theme
Fantastic Adventure

Use the Data

How many stars do you see?

What You Will Learn

In this chapter you will learn how to:

- Count, read, and write numbers to 50.
- Compare and order numbers to 50.
- Count by ones, fives, and tens to 100.
- Make a table to solve problems.

Dear Family,
In Chapter 8, I will learn about numbers to 100.
Here are vocabulary words and an activity that we can do together.

Counting Collections

- Give your child any number of objects up to 50 and ask him or her to count and tell how many there are.
- Now have your child use the cups and put the items into groups of 10.

- Ask how many groups of ten there are and how many ones.
- Say the number name.

use

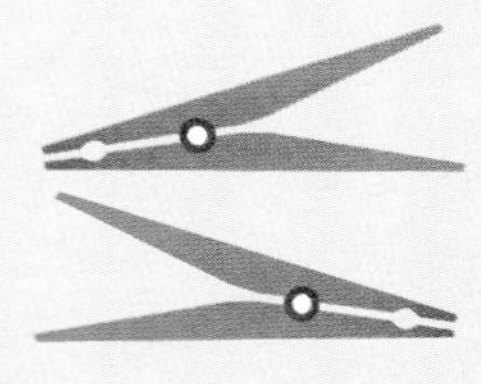

50 beans

or
50 clothespins

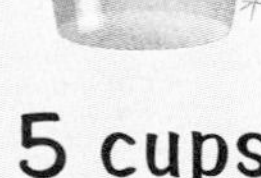

5 cups

Math Words

tens and ones

3 tens and 2 ones is 32.

order

These numbers are in order.
23, 24, 25

Additional activities at
www.mhschool.com/math

Name ____________________

2 tens 1 one 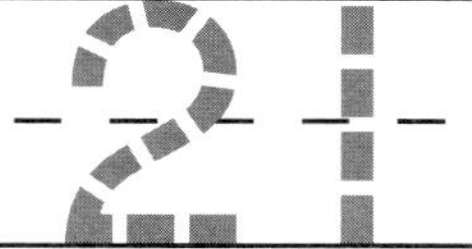

2 tens 2 ones  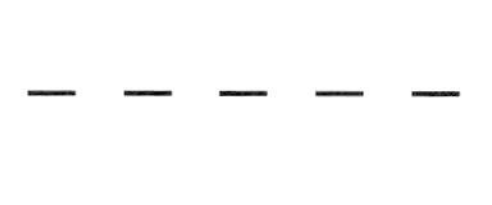

2 tens 3 ones 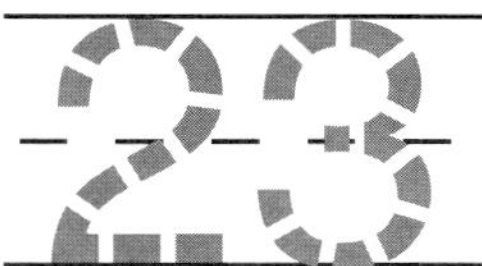

2 tens 4 ones 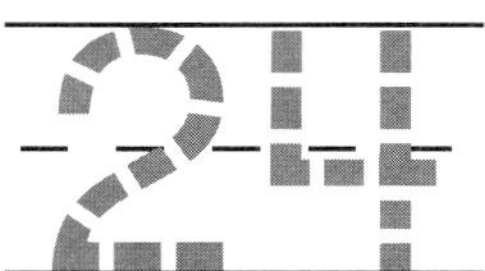 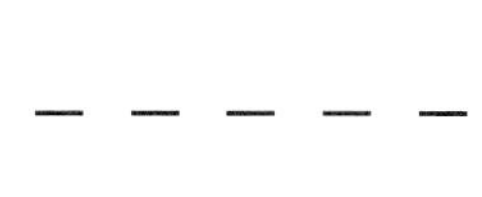

2 tens 5 ones 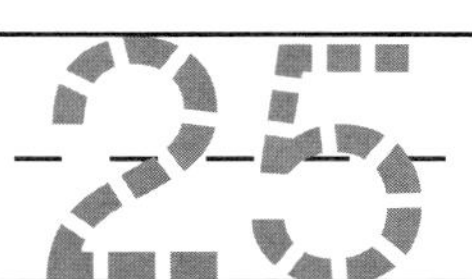

Use cubes and the 10-frames to show tens and ones. Write the number.

Math at Home: Your child made groups of 21 to 25 objects and wrote the numbers.
Activity: Show your child 21 to 25 small objects such as beans or paper clips. Have your child arrange the objects as groups of ten and ones.

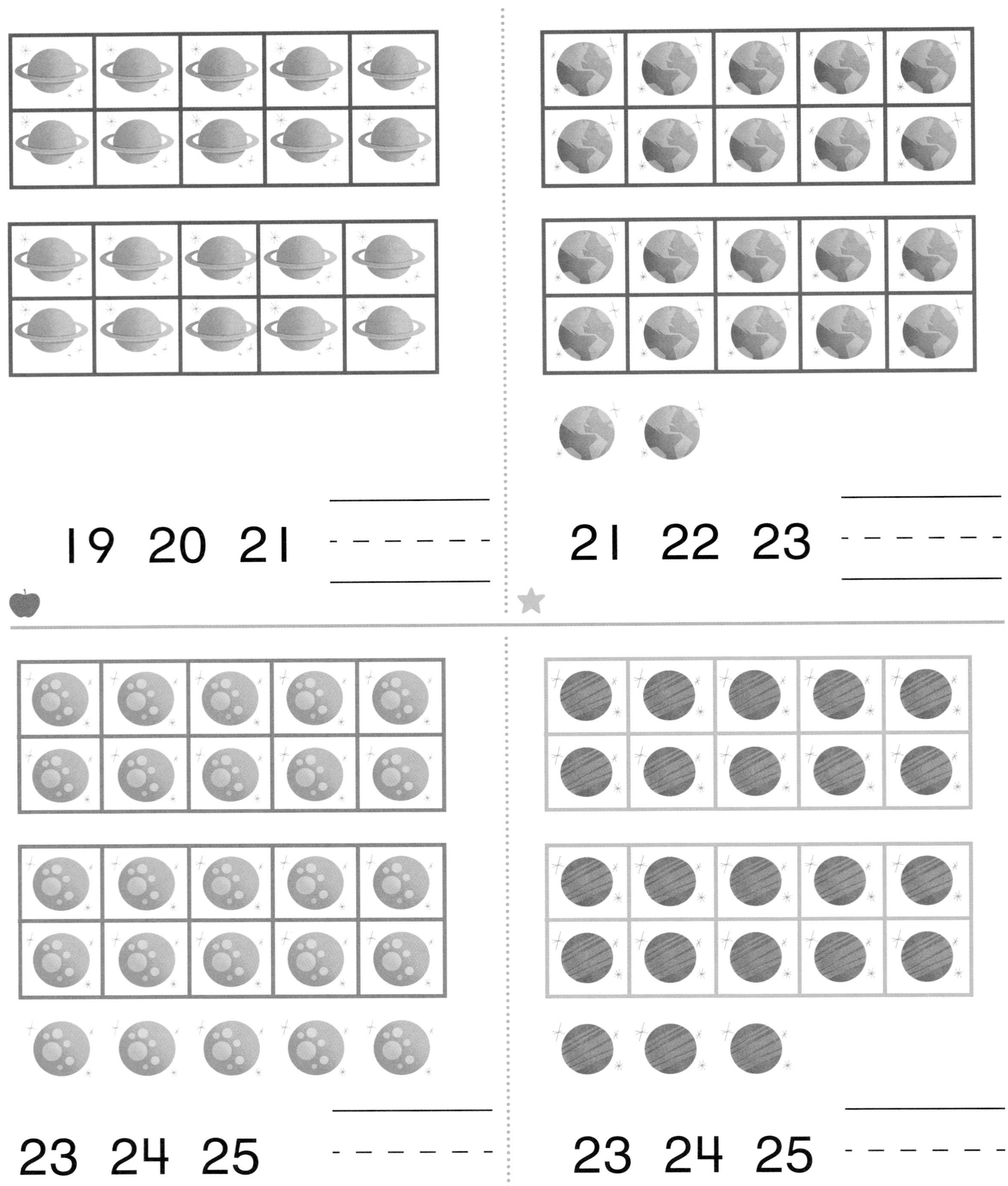

Count the objects. Circle the number. Write the number.

Name

2 tens 6 ones

26

2 tens 7 ones

27

2 tens 8 ones

28

2 tens 9 ones

29

3 tens 0 ones

30

Color to show ones. Write the number.

Math at Home: Your child colored to show the numbers 26 to 30.
Activity: Show your child 26 to 30 small objects such as beans or paper clips. Have your child arrange the objects as groups of ten and ones.

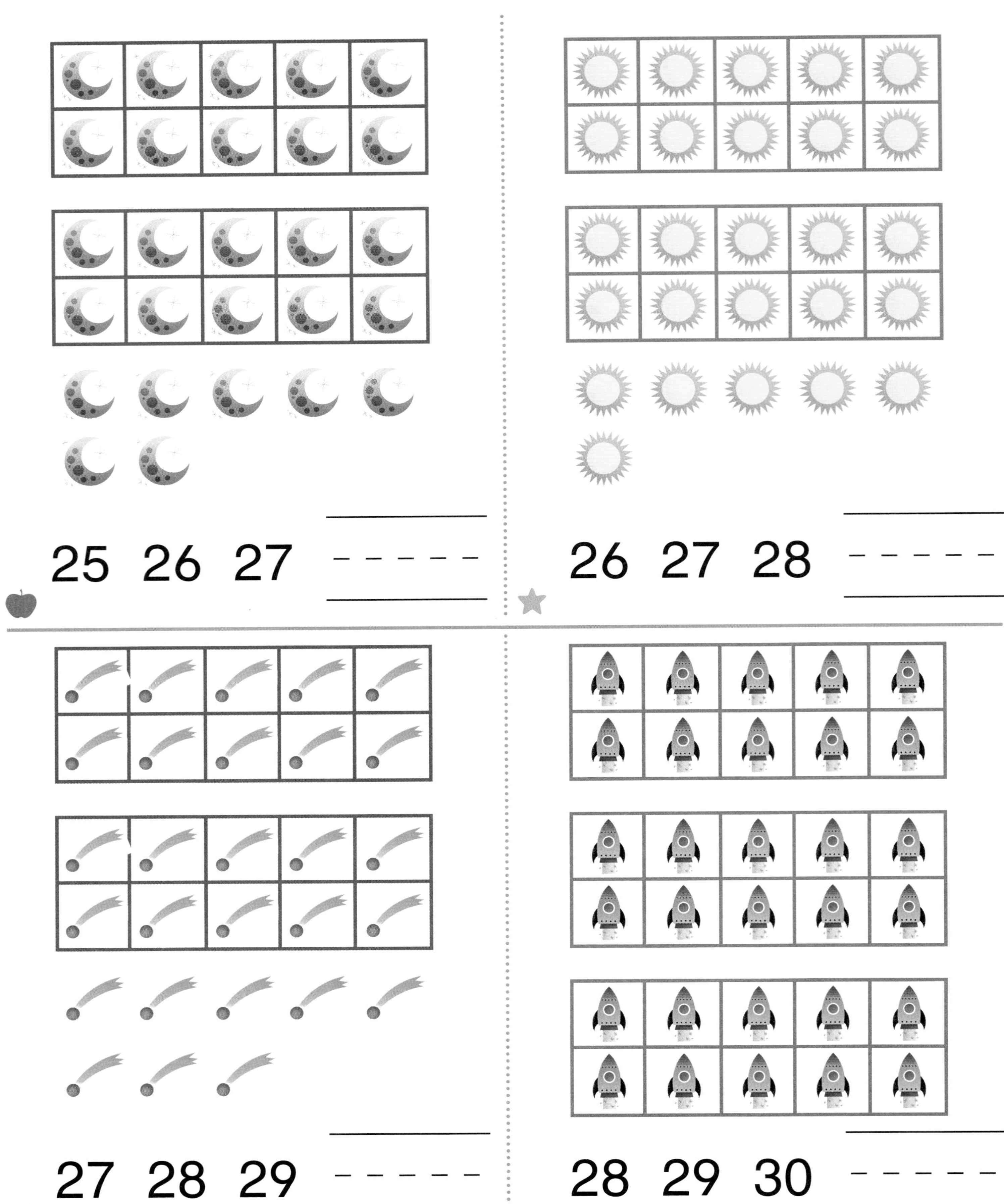

Count the objects. Circle the number. Write the number.

Name ______

tens	ones

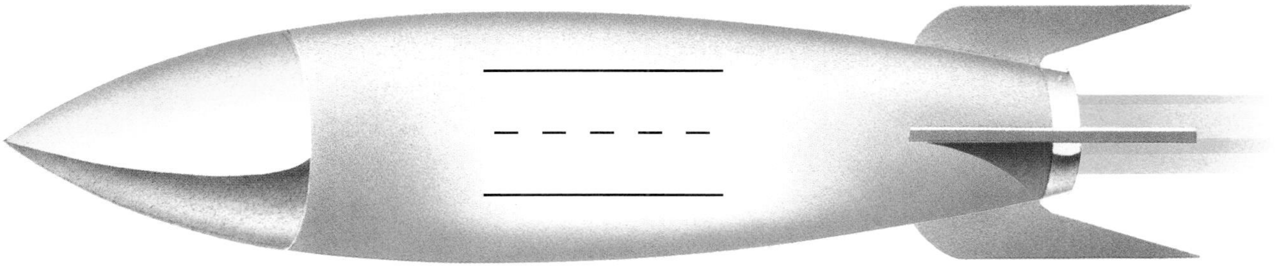

Use cubes to show numbers to 50. Draw cubes to show one number and then write the number in the box.

Math at Home: Your child made groups of up to 50 objects and wrote one number.
Activity: Show your child 30 to 50 small objects such as beans or paper clips. Have your child arrange the objects as groups of ten and ones.

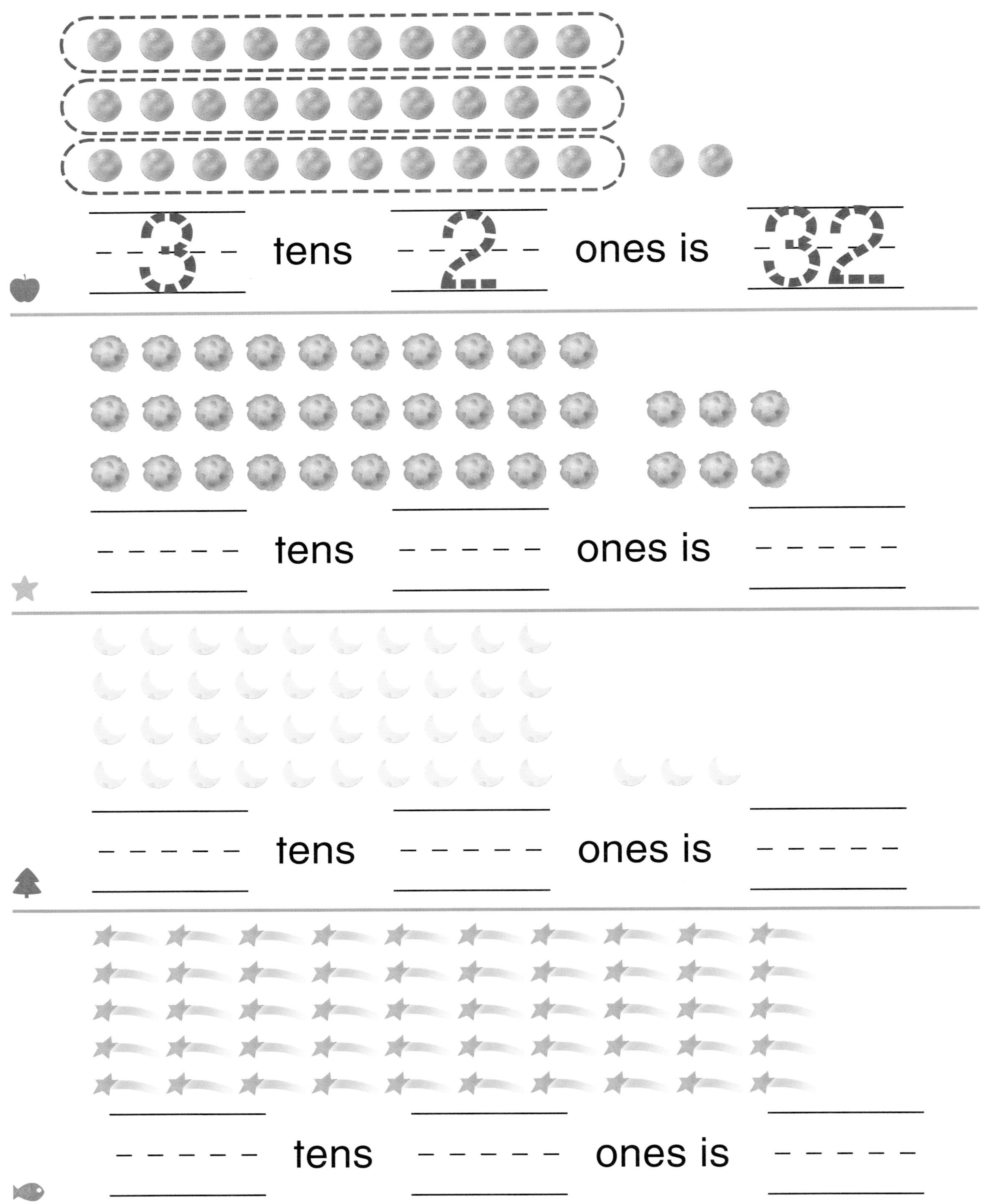

Circle groups of 10. Count and write the number.

Name ______________________

2 tens 3 ones

2 tens 4 ones

2 tens 8 ones

2 tens 7 ones

Use the 10 frames. Draw circles to show each number. Write the number.
Circle the number that is more.

Math at Home: Your child compared numbers to 50.
Activity: Ask your child to tell how he or she knows which numbers are more.

2 tens 5 ones

1 ten 7 ones

2 tens 4 ones

1 ten 8 ones

2 tens 9 ones

2 tens 0 ones

3 tens 0 ones

2 tens 1 one

Count the tens and ones and write each number. Circle the number that is fewer.

Name ____________________

Make a Table

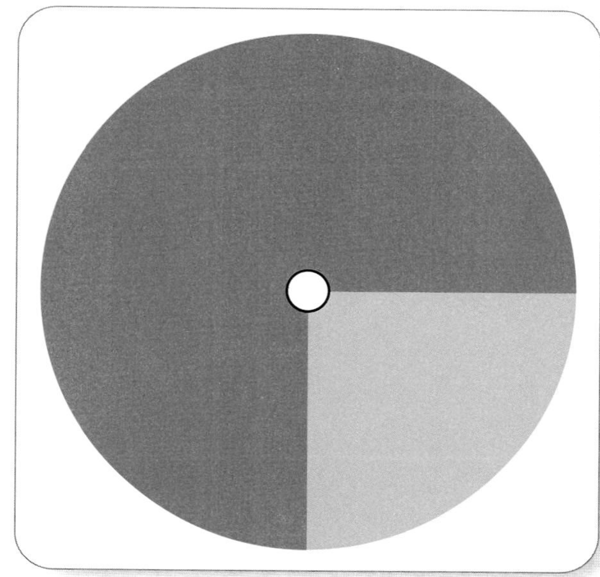

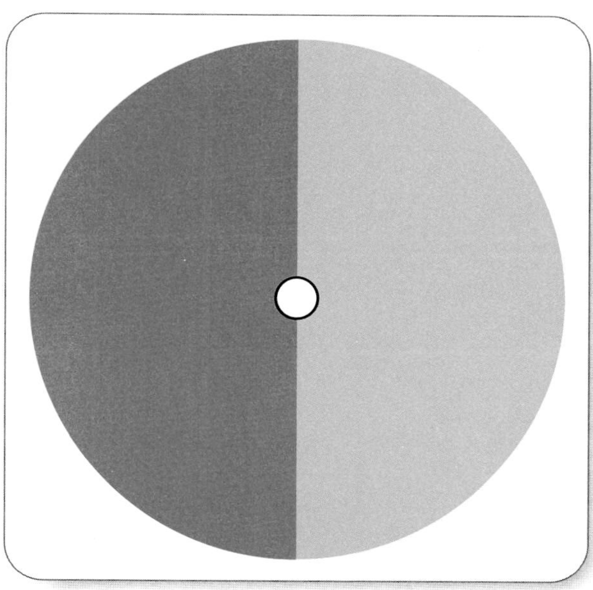

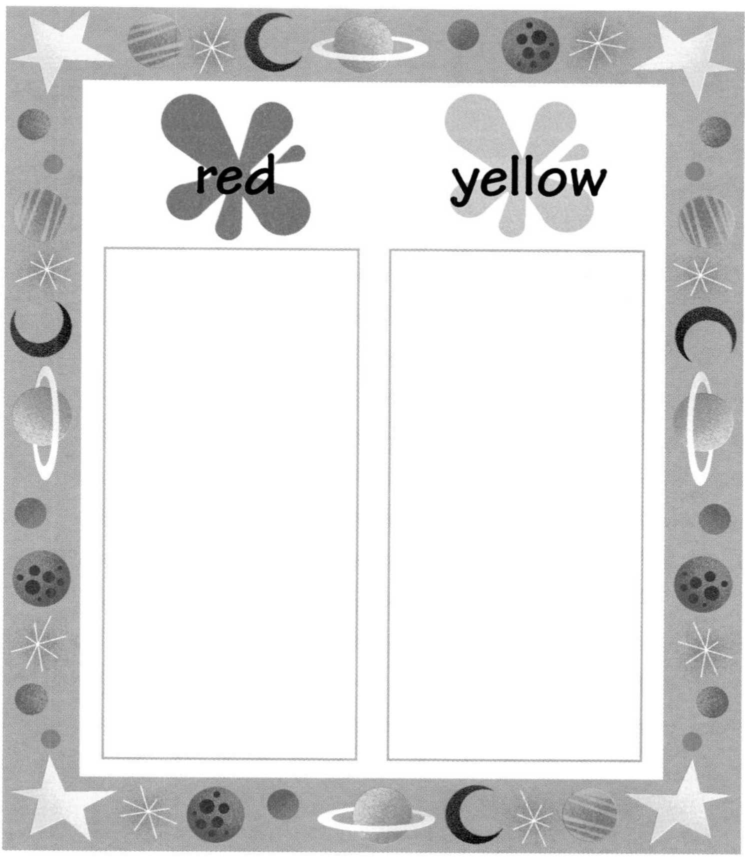

Use a pencil and a paper clip to make each spinner. Spin each spinner 10 times.
Use tallies to record your spins. Did you get more red or yellow?

Math at Home: Your child used tally marks and a spinner to learn about probability.
Activity: Ask your child to tell you what each tally mark shows.

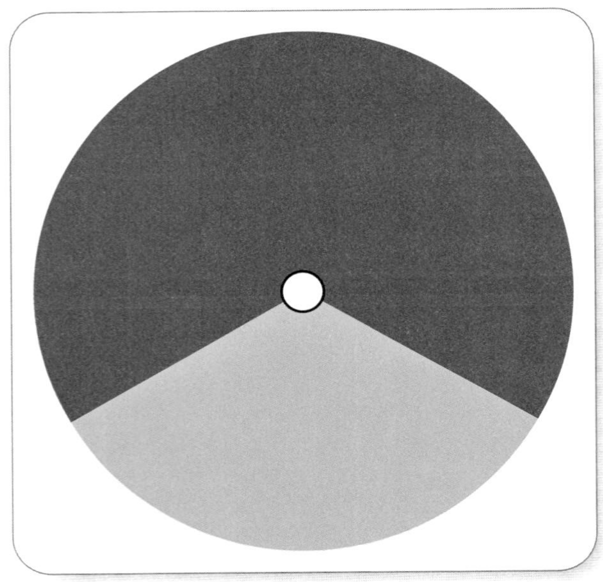

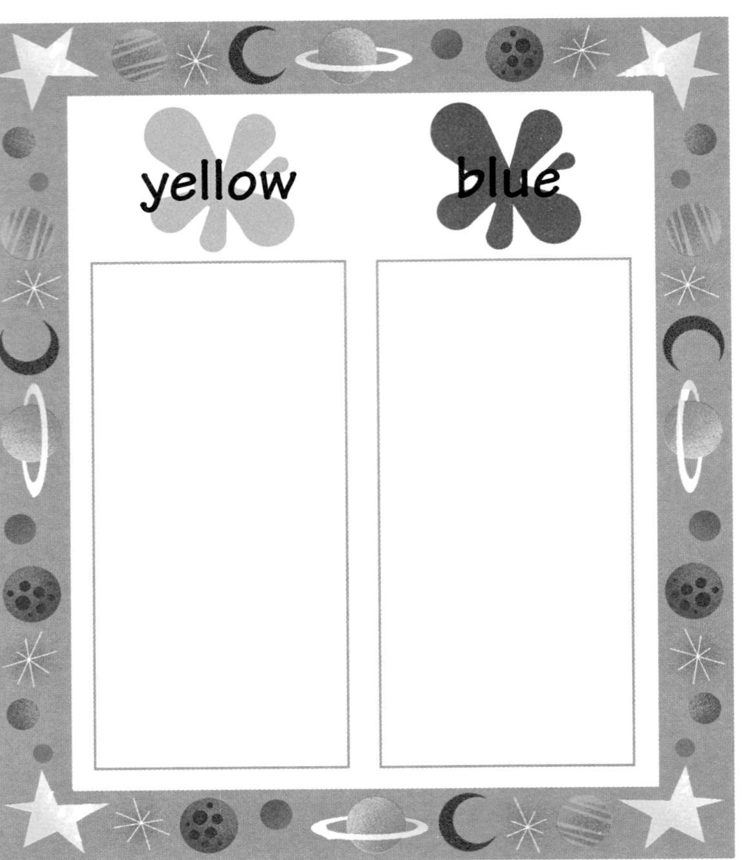

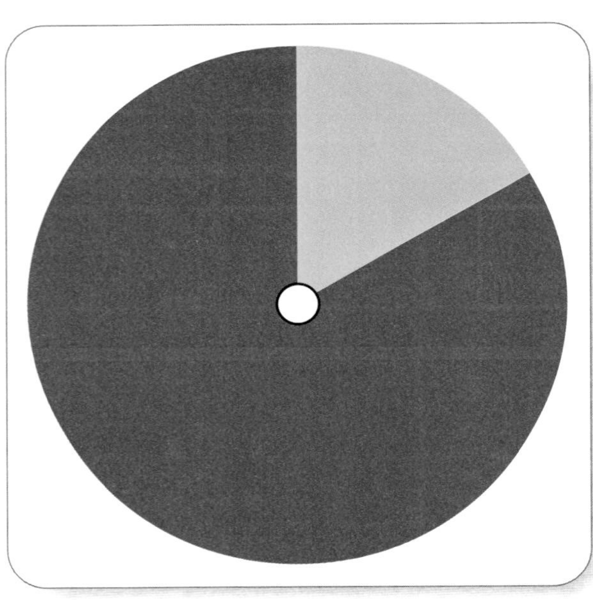

Look at each spinner. Which color are you more likely to spin? Circle the color name. Use a pencil and a paper clip to make each spinner. Spin each spinner 10 times. Use tallies to record your spins. Were you right?

Name ____________________

1	2	3	4	5	6	7	8	9	10
11	12	13	14	15	16	17	18	19	20
21	22	23	24	25	26	27	28	29	30
31	32	33	34	35	36	37	38	39	40
41	42	43	44	45	46	47	48	49	50
51	52	53	54	55	56	57	58	59	60
61	62	63	64	65	66	67	68	69	70
71	72	73	74	75	76	77	78	79	80
81	82	83	84	85	86	87	88	89	90
91	92	93	94	95	96	97	98	99	100

Before		After
____	12	____
____	44	____
____	56	____
____	97	____

Find the number on the hundred chart. Color the number just before red. Write the number. Color the number just after yellow. Write the number.

Math at Home: Your child learned about the order of the numbers 1 to 100.
Activity: Say a number. Ask your child to find the number on the chart and say the numbers that come before and after.

1	2	3	4	5	6	7	8	9	10
11	12	13	14	15	16	17	18	19	20
21	22	23	24	25	26	27	28	29	30
31	32	33	34	35	36	37	38	39	40
41	42	43	44	45	46	47	48	49	50
51	52	53	54	55	56	57	58	59	60
61	62	63	64	65	66	67	68	69	70
71	72	73	74	75	76	77	78	79	80
81	82	83	84	85	86	87	88	89	90
91	92	93	94	95	96	97	98	99	100

Before		After
_____	14	_____
_____	28	_____
_____	43	_____
_____	77	_____
_____	89	_____

Find each number on the hundred chart.

Write the number that comes just before. ★ Write the number that comes just after.

Name ____________________

Count by Tens

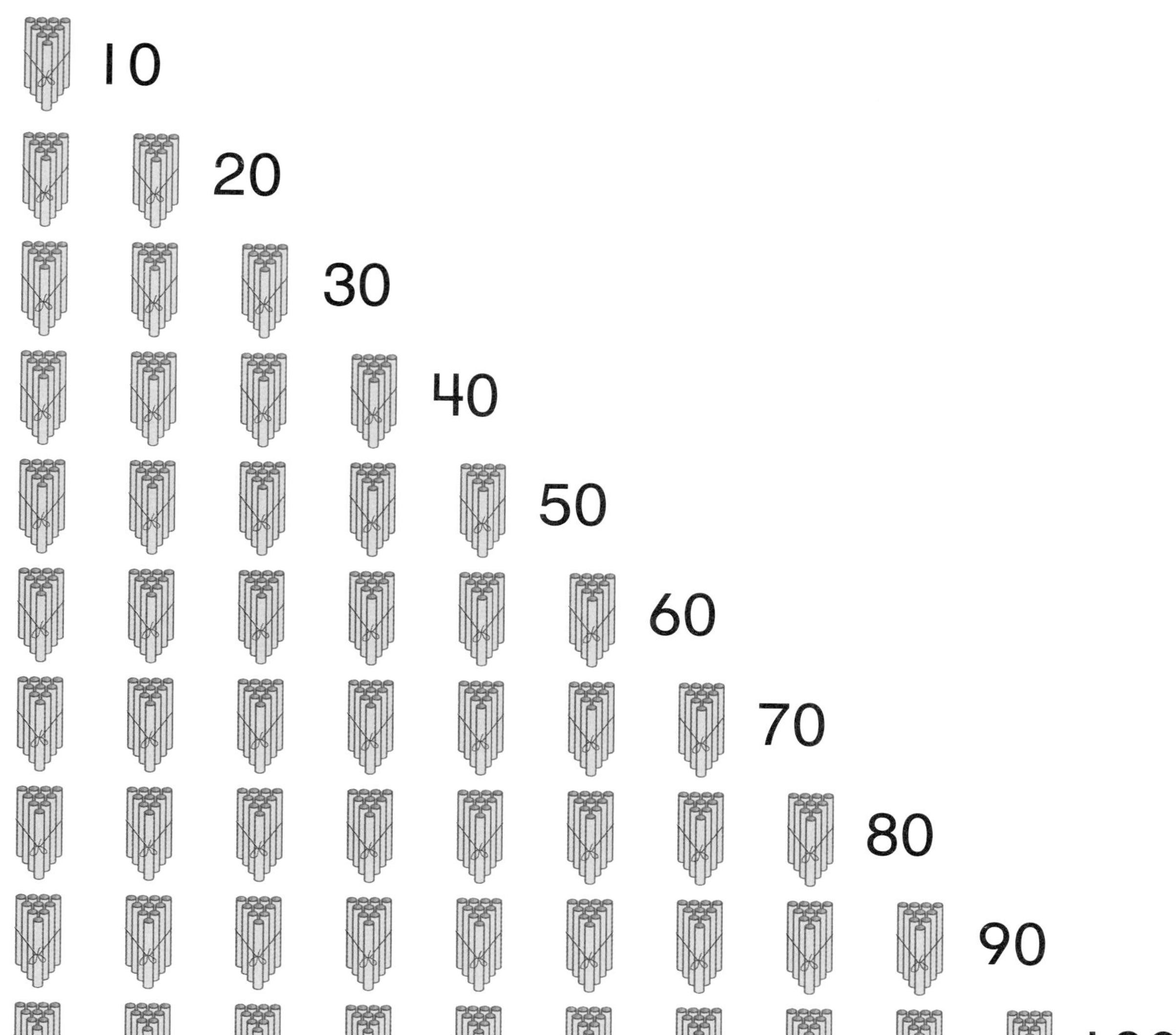

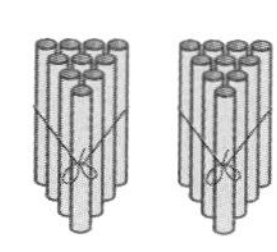

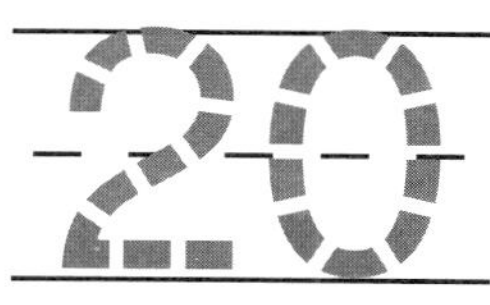

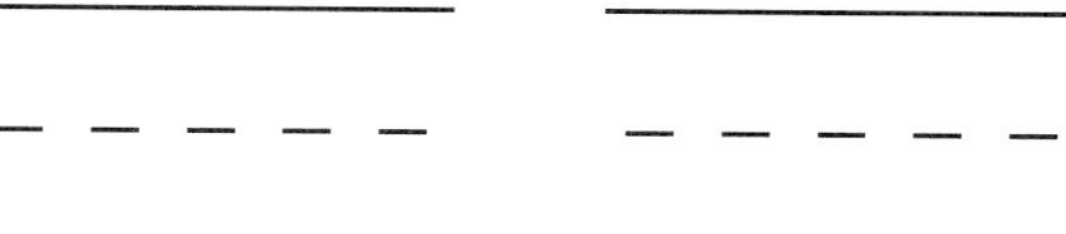

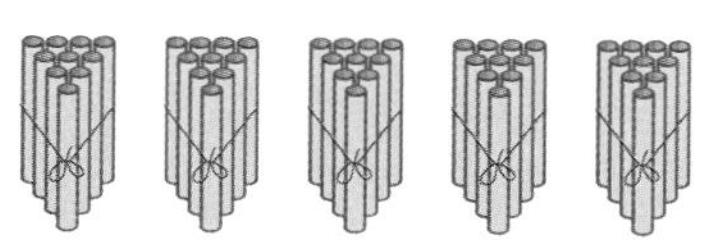

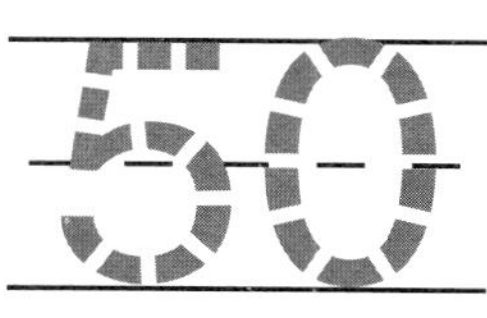

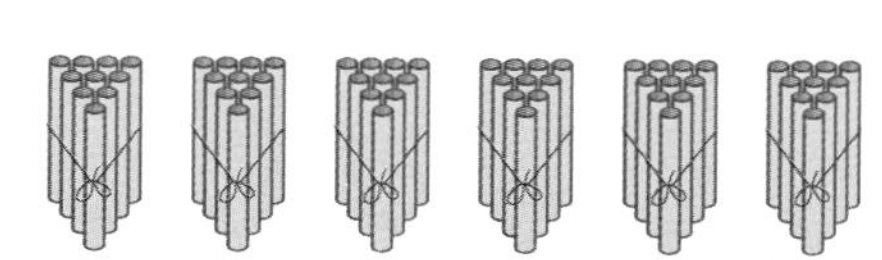

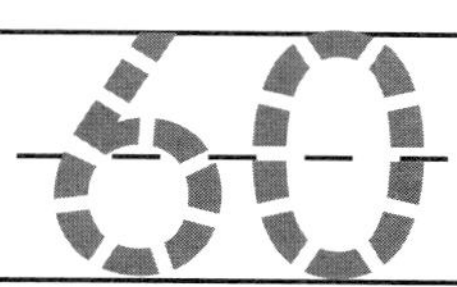

Count by tens. Write the number.

Math at Home: Your child counted by tens to 100.
Activity: Have your child arrange small objects in groups of ten and tell how many there are in all.

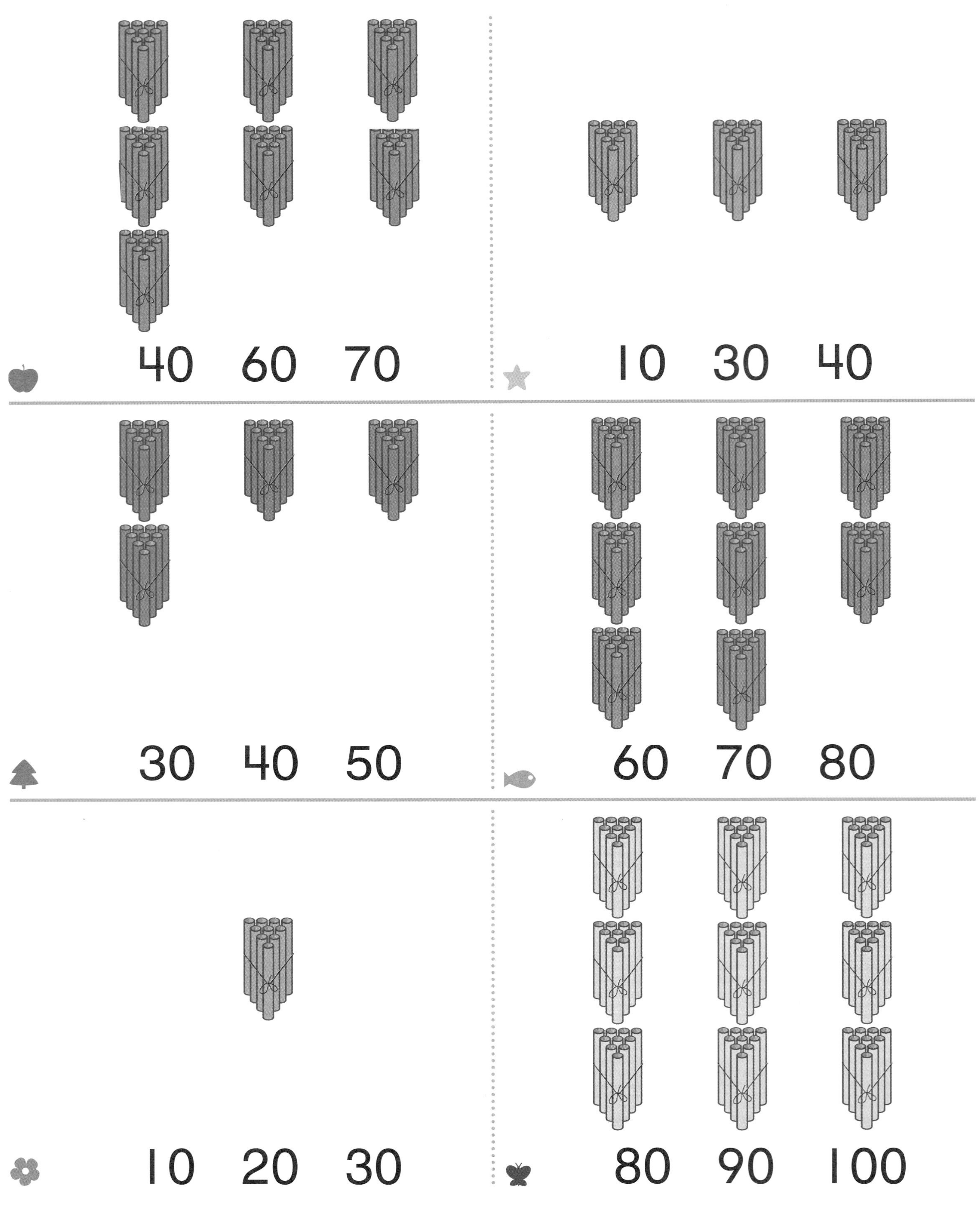

Count by tens. Circle the number.

Name____________________

Put 5 cubes in each box. Count by fives. Write the missing numbers.

Math at Home: Your child counted by fives to 100.
Activity: Have your child arrange small objects such as pennies in groups of five and tell how many there are in all.

1	2	3	4	5	6	7	8	9	10
11	12	13	14	15	16	17	18	19	20
21	22	23	24	25	26	27	28	29	30
31	32	33	34	35	36	37	38	39	40
41	42	43	44	45	46	47	48	49	50
51	52	53	54	55	56	57	58	59	60
61	62	63	64	65	66	67	68	69	70
71	72	73	74	75	76	77	78	79	80
81	82	83	84	85	86	87	88	89	90
91	92	93	94	95	96	97	98	99	100

Count by fives. Color the boxes yellow.
Count by tens. Use red to circle the numbers.

Name ______________________________

About How Many?	Count	Write
5 10 100		______
5 10 100		______

Take a handful of counters. Circle about how many.
Make stacks of 5 or 10 to count. Write how many.

Math at Home: Your child told about how many items were in a group, then counted.
Activity: Have your child look at a group of items at home, such as books on a shelf or buttons in a button box, and tell about how many there are.

5 20 100

5 20 100

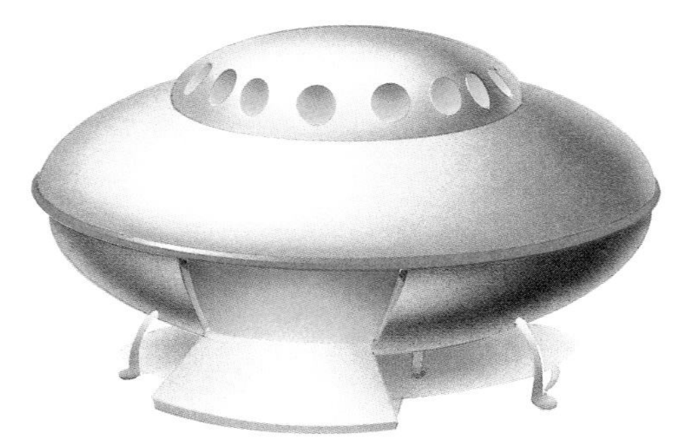

5 20 100

5 20 100

Circle to tell about how many.

Name ______________________________

22 23 24 | 27 28 29

______ tens ______ ones is ______

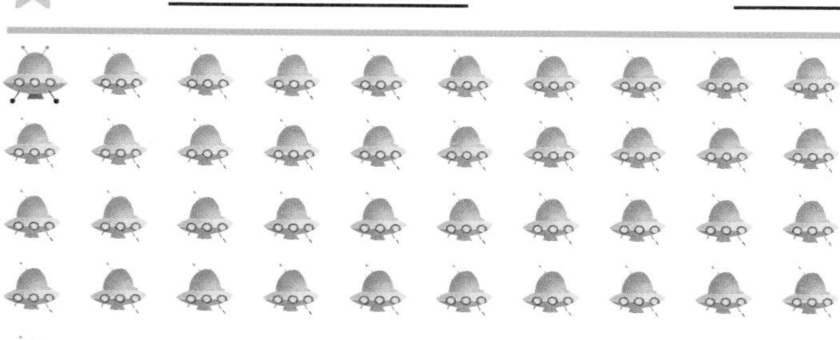

______ tens ______ ones is ______

______ | ______

McGraw-Hill School Division

Count the objects. Circle the number.
Circle groups of ten. Count and write the number.
Count the tens and ones and write each number. Circle the number that is more.

31	32	33	34	35	36	37	38	39	40
41	42	43	44	45	46	47	48	49	50
51	52	53	54	55	56	57	58	59	60
61	62	63	64	65	66	67	68	69	70
71	72	73	74	75	76	77	78	79	80

Before		After
______	32	______
______	48	______
______	65	______

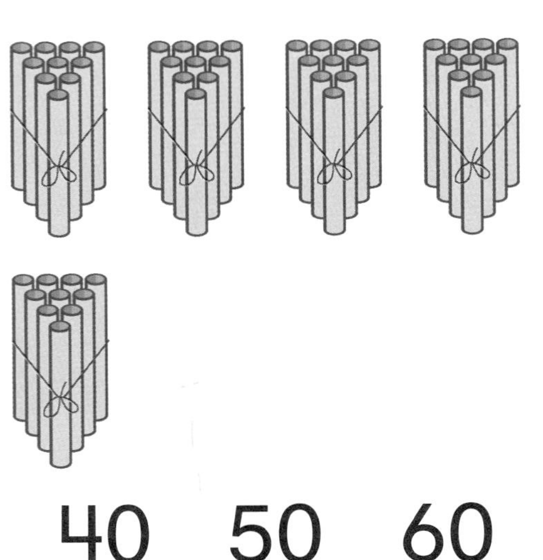

40 50 60

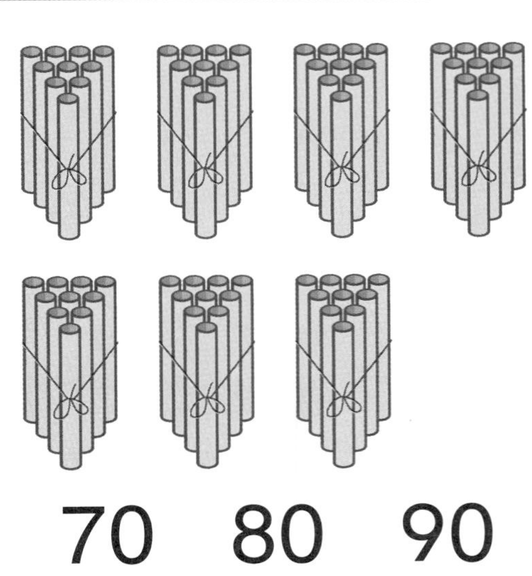

70 80 90

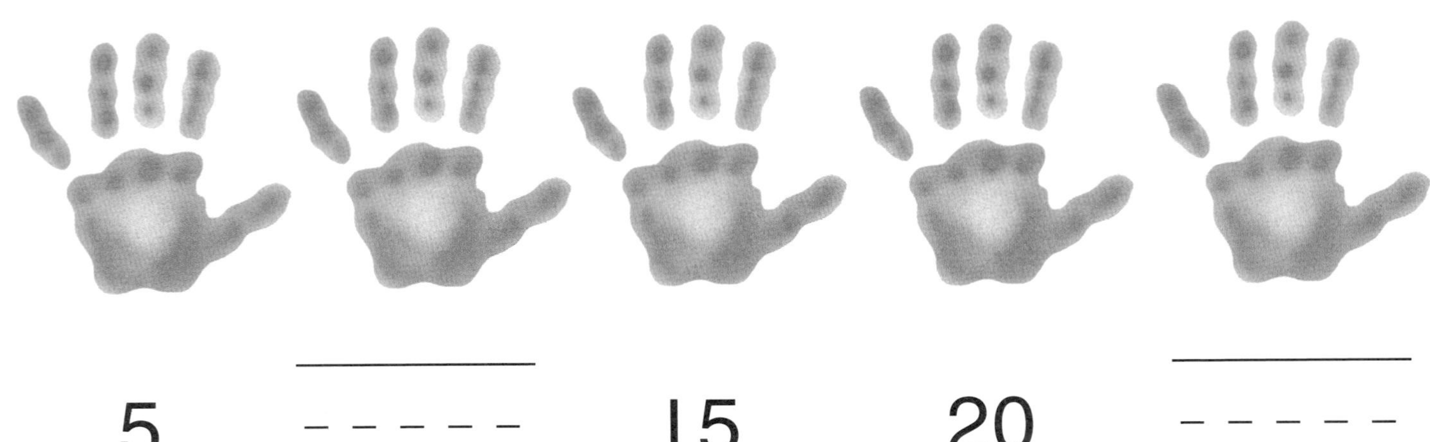

5 ______ 15 20 ______

Find the number on the number chart. Color the number just before red. Write the number. Color the number just after yellow. Write the number. Count by tens. Circle the number. Count by fives. Write the missing numbers.

CHAPTER 9

Money

theme
Going Shopping

Use the Data

How are the children using money?

What You Will Learn

In this chapter you will learn how to:

- Identify pennies, nickels, dimes, quarters and dollar bills.
- Identify the value of pennies, nickels, and dimes.
- Act out situations to solve problems.

Dear Family,
In Chapter 9, I will learn about money. Here are some new vocabulary words and an activity that we can do together.

Kitchen Store

- Write prices less than 20¢ on sticky notes and attach these prices to kitchen items.

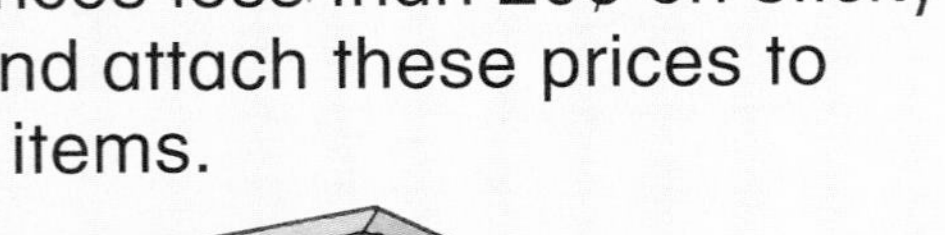

- Have your child use pennies to "buy" different items.
- You can extend the activity by allowing your child to make small purchases at a store or at garage sales.

use

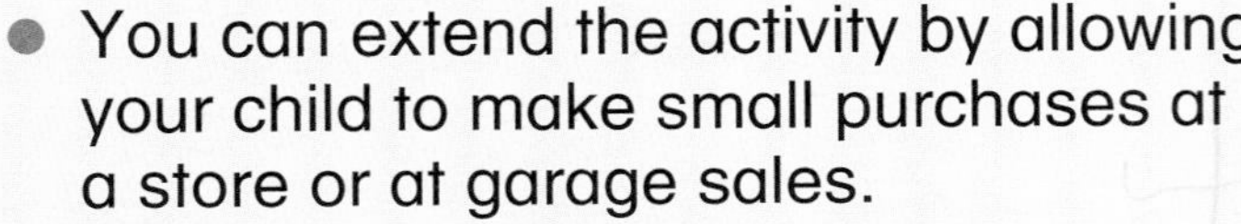

kitchen items

sticky notes

pennies

Math Words

penny
one cent
1¢

nickel
five cents
5¢

dime
ten cents
10¢

quarter
twenty-five cents
25¢

dollar
one dollar
$1.00

Additional activities at
www.mhschool.com/math

Name ______________________________

1¢ 1¢

Math Words
penny
cent (¢)

Say these words.

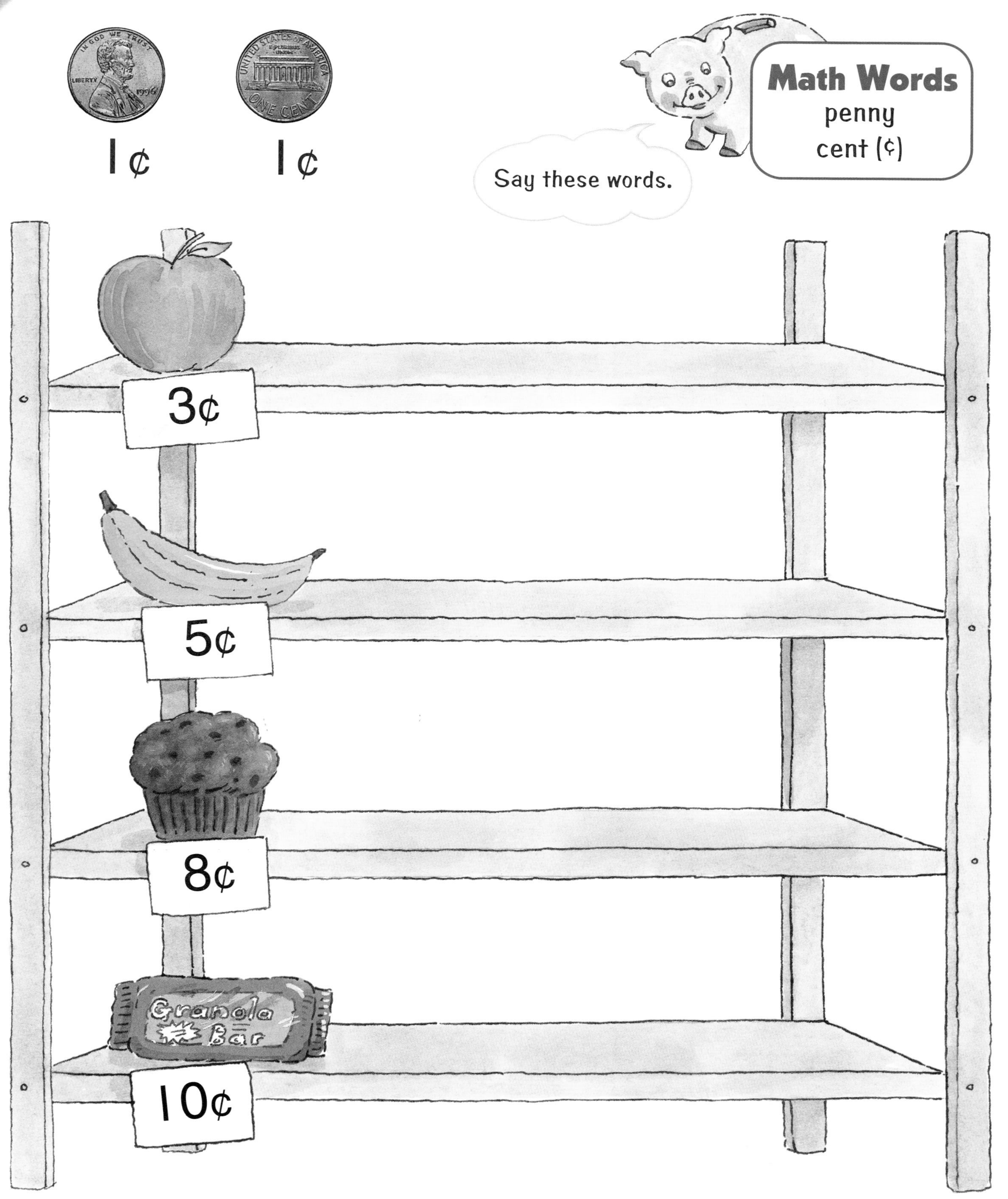

Use pennies to show the money amounts. Draw the pennies.

Math at Home: Your child is learning about pennies.
Activity: Give your child 10 pennies. Ask him or her to show you different amounts of money from 1¢ to 10¢.

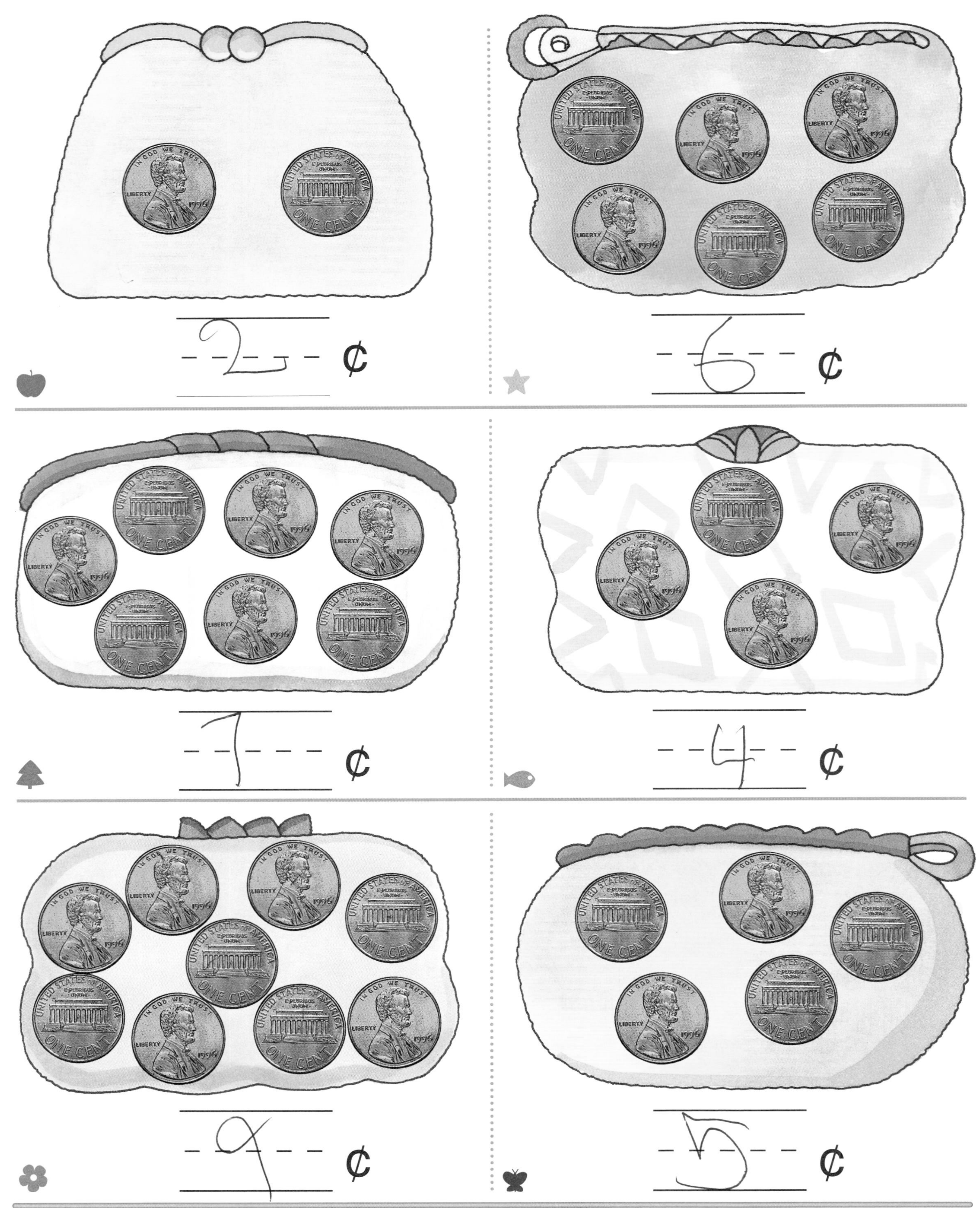

Count the pennies. Write the amount.

Name________________________________

5¢ 5¢

Math Word
nickel

Say this word.

10 ¢

Circle the nickels. Count. Write the amount.

Math at Home: Your child is learning about nickels.
Activity: Give your child a group of pennies and nickels. Ask your child to show you which ones are nickels.

3¢ 10¢ 17¢

4¢ 20¢ 25¢

6¢ 25¢ 30¢

2¢ 5¢ 13¢

Count the coins. Circle the amount.

Name ________________________________

Math Word
dime

10¢

10¢

Say this word.

50 ¢

Circle the dimes. Count. Write the amount.

Math at Home: Your child is learning about dimes.
Activity: Give your child a handful of dimes, nickels and pennies. Ask your child to show you which ones are dimes.

10¢ 20¢ 35¢

4¢ 41¢ 60¢

10¢ 32¢ 40¢

30¢ 40¢ 50¢

Count the coins. Circle the amount.

Name ____________________

1¢ 5¢ 10¢

____ ¢

26¢ 31¢ 24¢ 47¢ 40¢

Count the coins. Write the amount. Circle the toy you can buy.

Math at Home: Your child practiced using pennies, nickels and dimes.
Activity: Give your child some dimes, nickels and pennies, up to .25¢. Ask your child to count the coins and draw a picture of what he or she would buy.

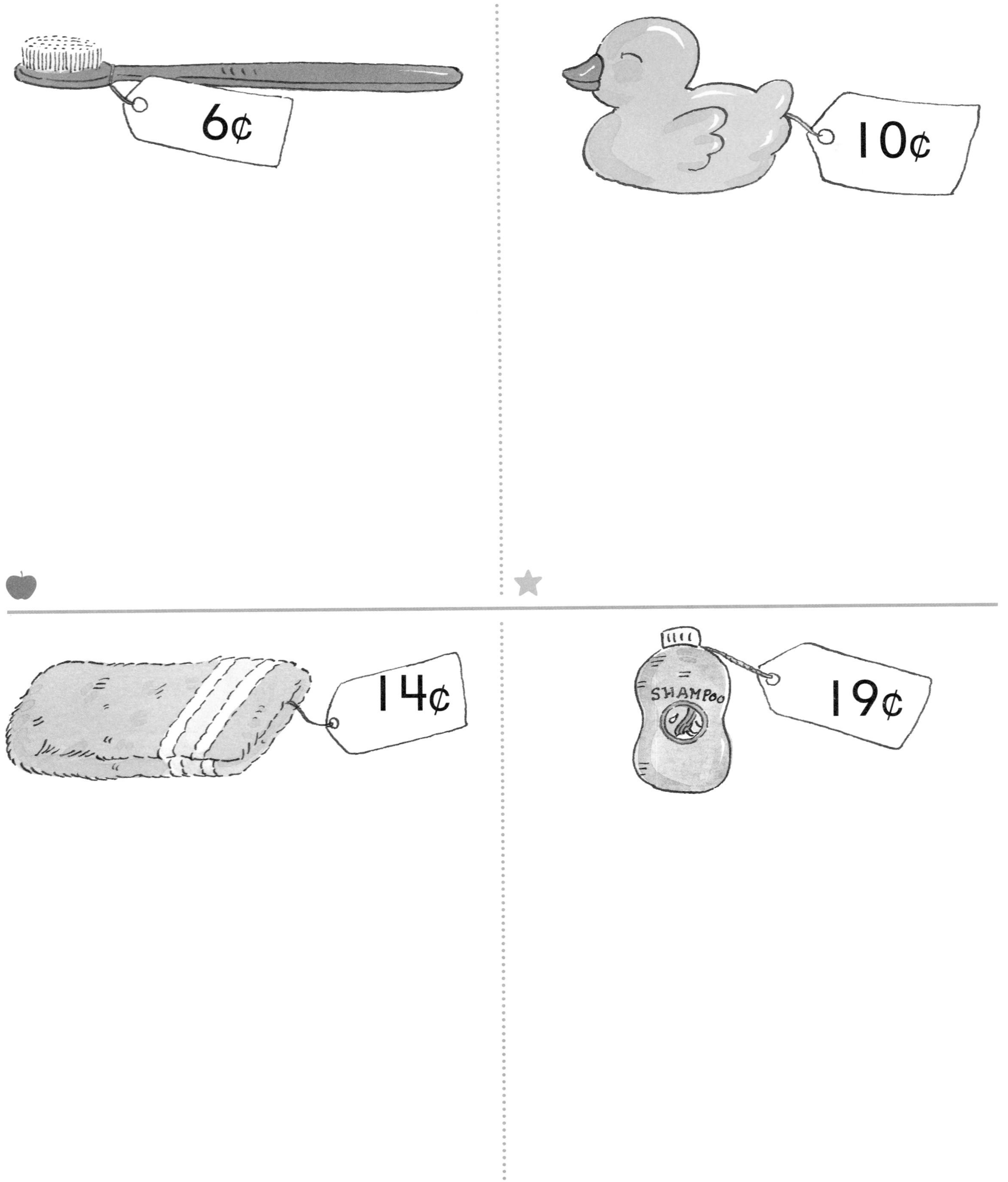

Use coins to show the money amount. Draw the coins.

Name ____________________

Act It Out

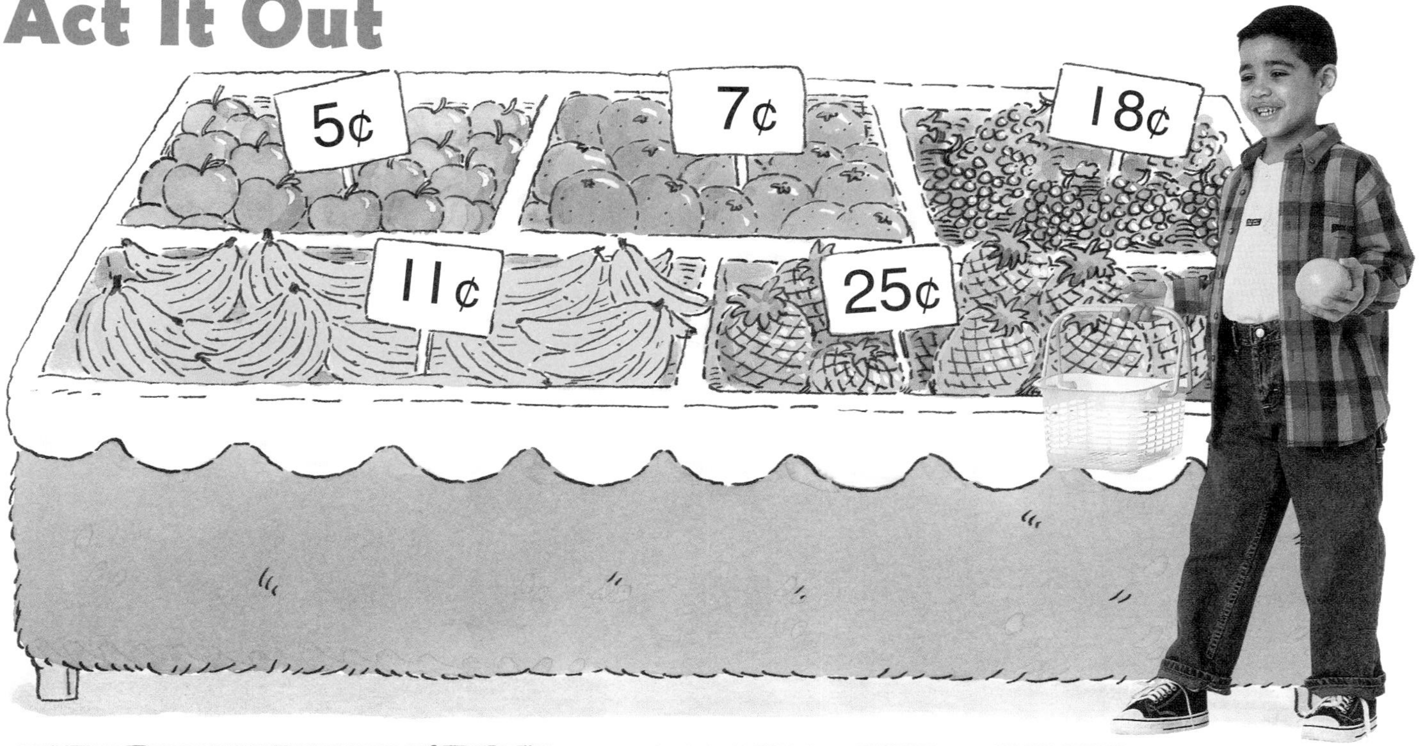

5¢ 7¢ 18¢
11¢ 25¢

		____ ¢
		____ ¢

Choose something you want to buy. Draw the item. Show the coins you can use. Write the amount.

Math at Home: Your child acted out buying items.
Activity: Give your child some pennies, nickels, and dimes. Ask him or her to act out buying something at home.

		______ ¢
		______ ¢

Chose something you want to buy. Draw the item. Show the coins you can use. Write the amount.

Name ____________________

25¢

$1.00

Math Words
quarter
dollar

Circle each quarter. Draw a line under each dollar bill.

Math at Home: Your child identified quarters and dollar bills.
Activity: Display some dollar bills and a variety of coins. Ask your child to identify the dollars and the quarters.

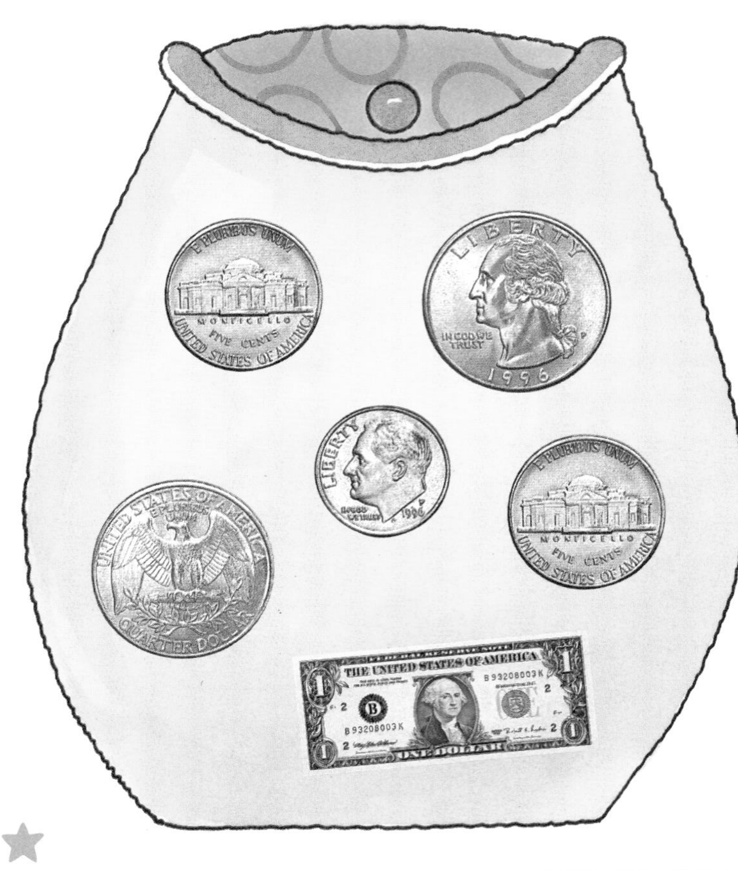

Circle each quarter. Draw a line under each dollar bill.

Name ______________________

______ ¢

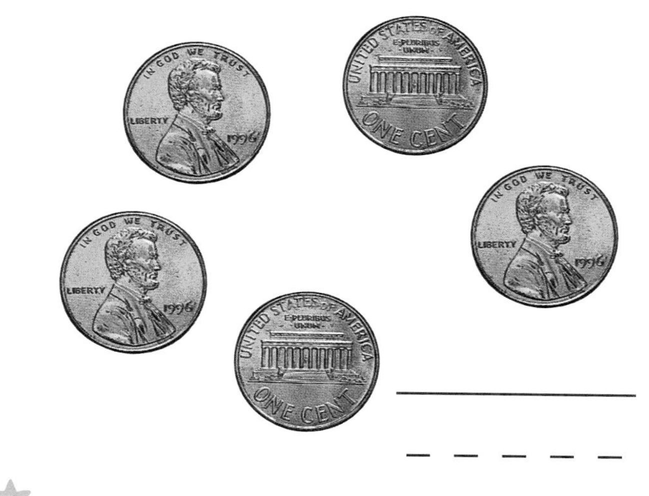

______ ¢

______ ¢

______ ¢

8¢ 21¢ 41¢

32¢ 70¢ 80¢

Count the pennies. Write the amount.

Circle the nickels. Count. Write the amount.

Circle the dimes. Count. Write the amount.

Count the coins. Circle the amount.

¢

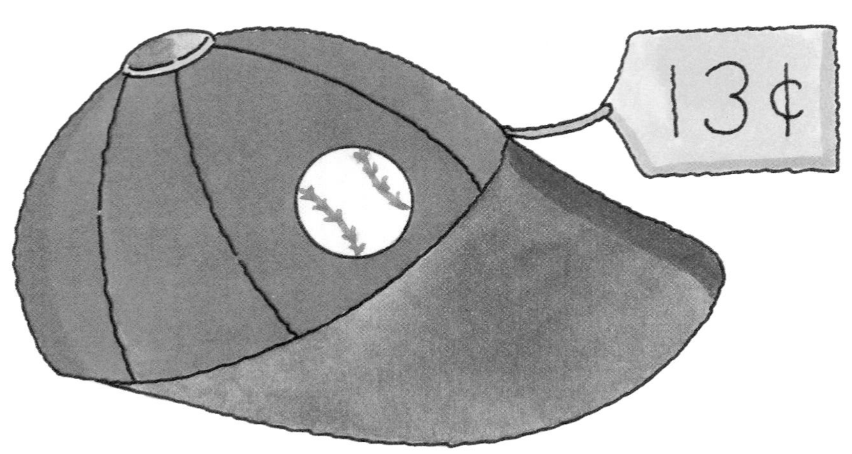

- Count the coins. Write the amount. Circle what you can buy.
- Show the coins you can use to buy the hat.
- Circle each quarter. Draw a line under each dollar bill.

CHAPTER 10

Measurement

theme

Construction Site

Use the Data

What are the beavers trying to find out about the log?

What You Will Learn

In this chapter you will learn how to:

- Find the longer and shorter of two objects.
- Measure length using nonstandard units.
- Explore capacity, weight, and temperature.
- Use guess and check to solve problems.

Dear Family,

In Chapter 10, I will learn about measurement.
Here are some new vocabulary words and an activity that we can do together.

Measurement Fun

Math Words

shorter | longer

lighter | heavier

hot | cold

- Give your child various sized items such as pasta, paper clips or toothpicks.
- Have your child use the objects to measure the lengths of different objects in your home.

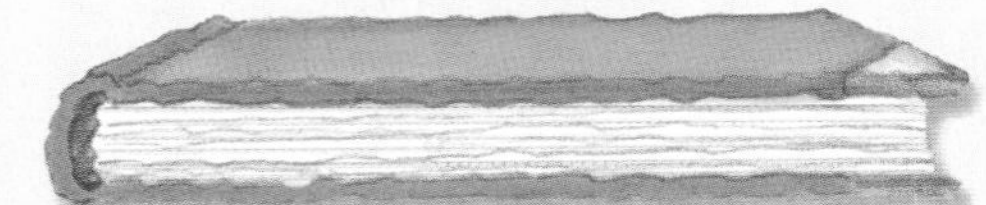

- Compare the measurements of two or three objects. Which is longer? shorter?

use

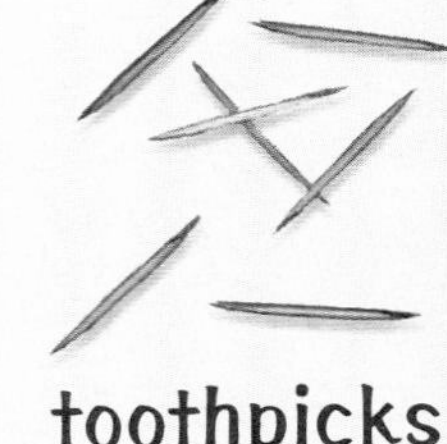

toothpicks

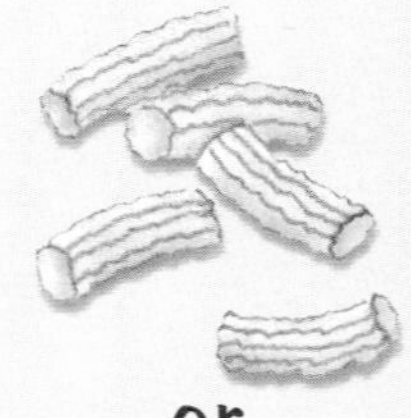

or
pasta

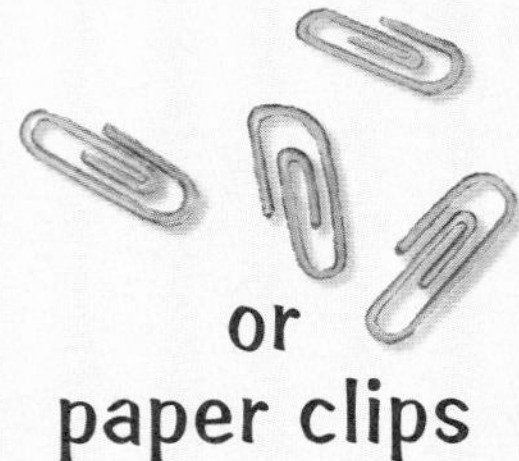

or
paper clips

Additional activities at
www.mhschool.com/math

Name________________________________

Say these words.

Math Words
longer
shorter

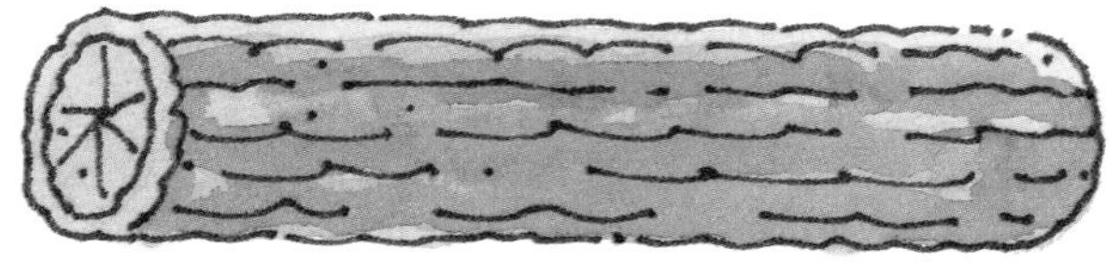

shorter

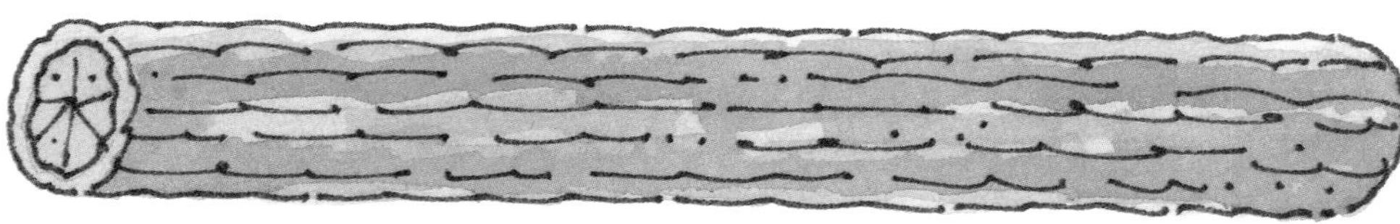

longer

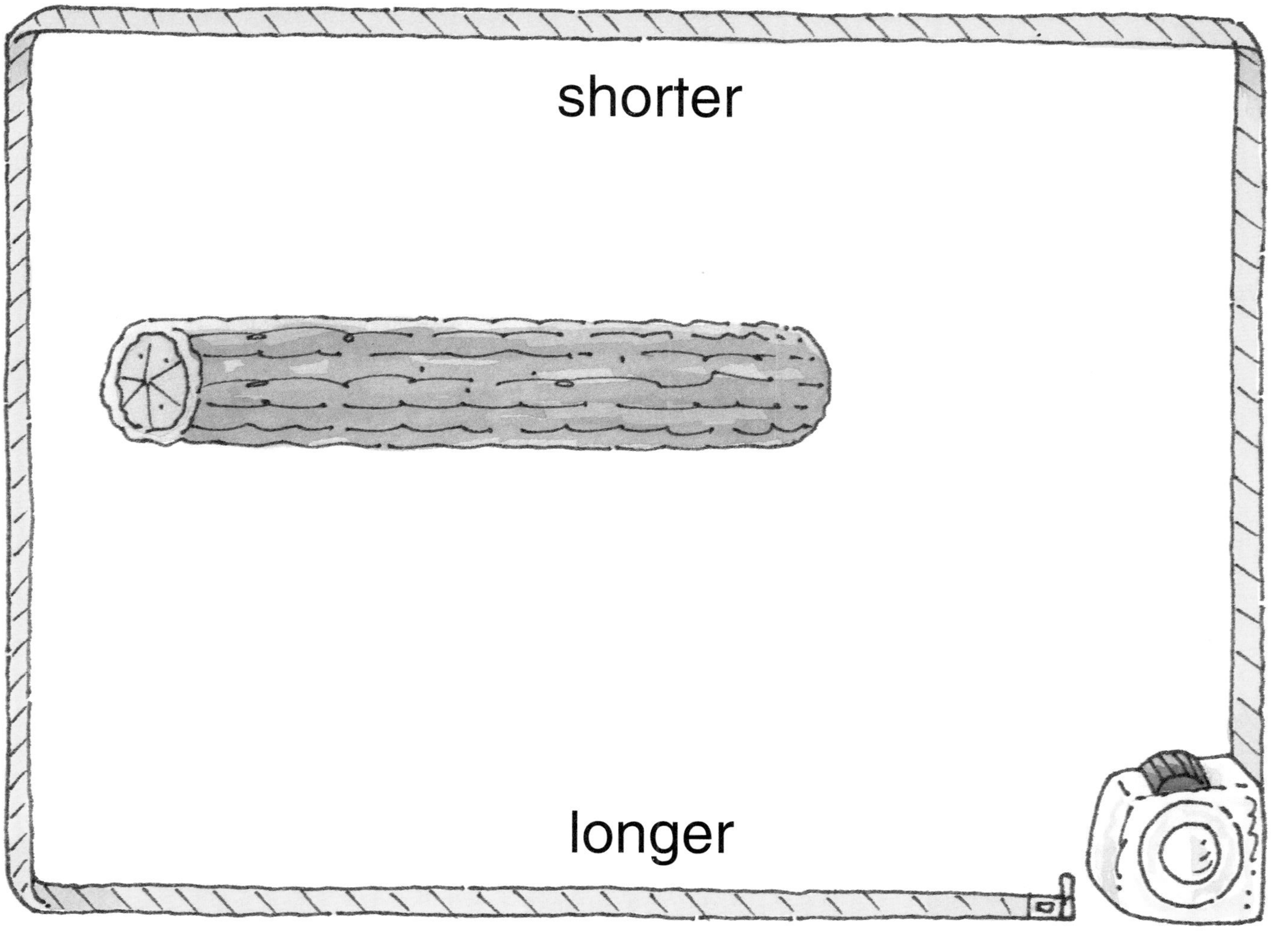

shorter

longer

Find an object that is shorter than the log. Draw it.
Find an object that is longer than the log. Draw it.

Math at Home: Your child is learning about longer and shorter.
Activity: Show your child a pencil. Ask him or her to find objects at home that are shorter than the pencil and objects that are longer than the pencil.

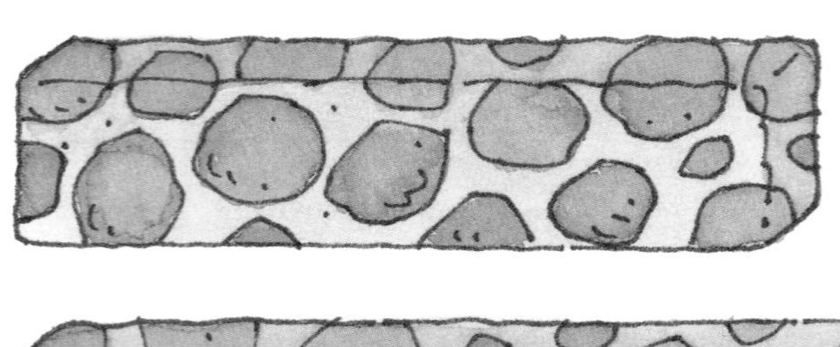

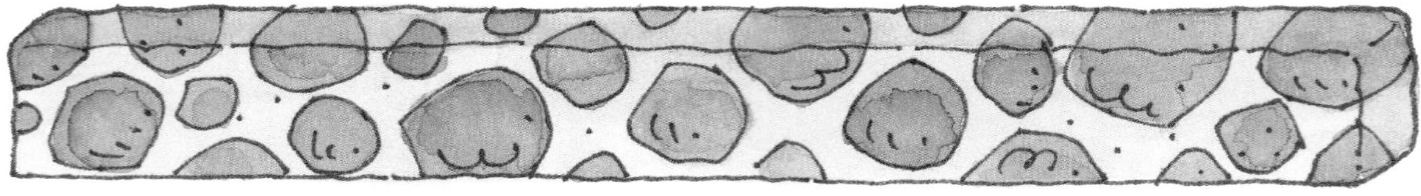

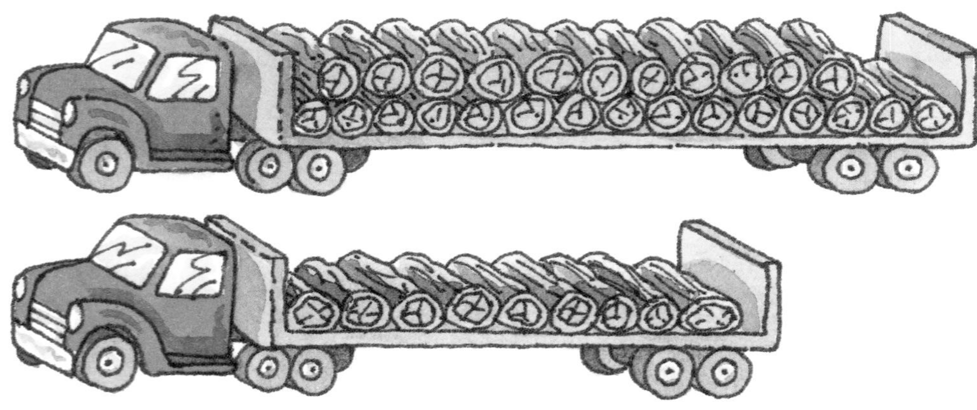

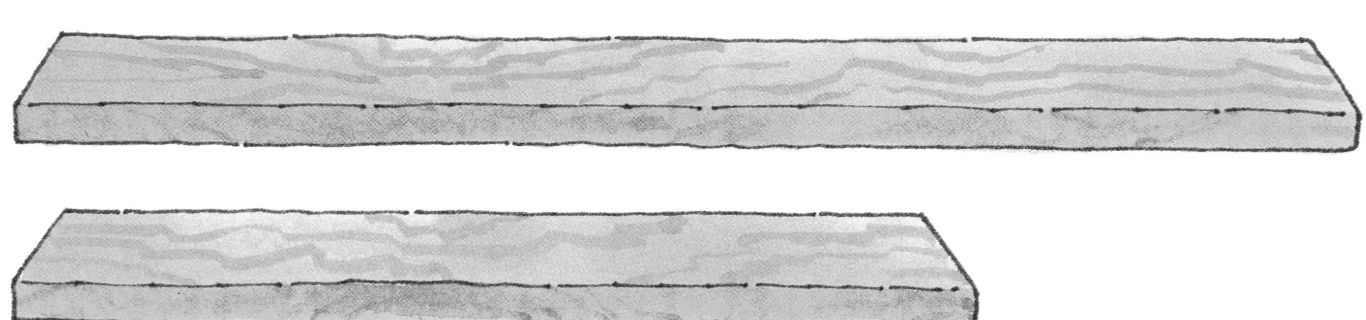

Circle the longer object.

Draw an X on the shorter object.

Name ____________________

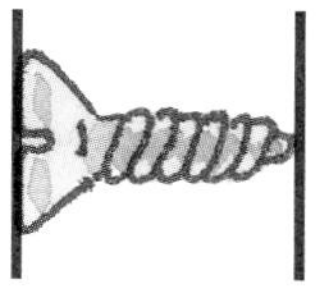

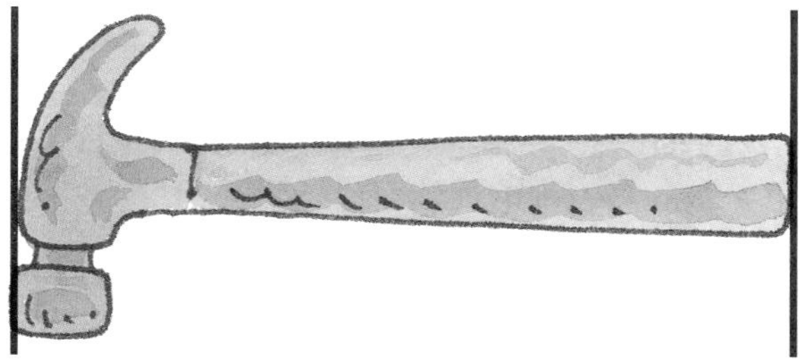

Use cubes to measure each object. Write how many.

Math at Home: Your child used cubes to measure.
Activity: Have your child use small items such as paper clips or pieces of pasta. Ask your child to find something that is as long as 5 paper clips.

About

About

About

Look at each object. Guess about how many cubes long. Write your guess.
Use cubes to measure each object. Write the number.

Name____

Guess and Check

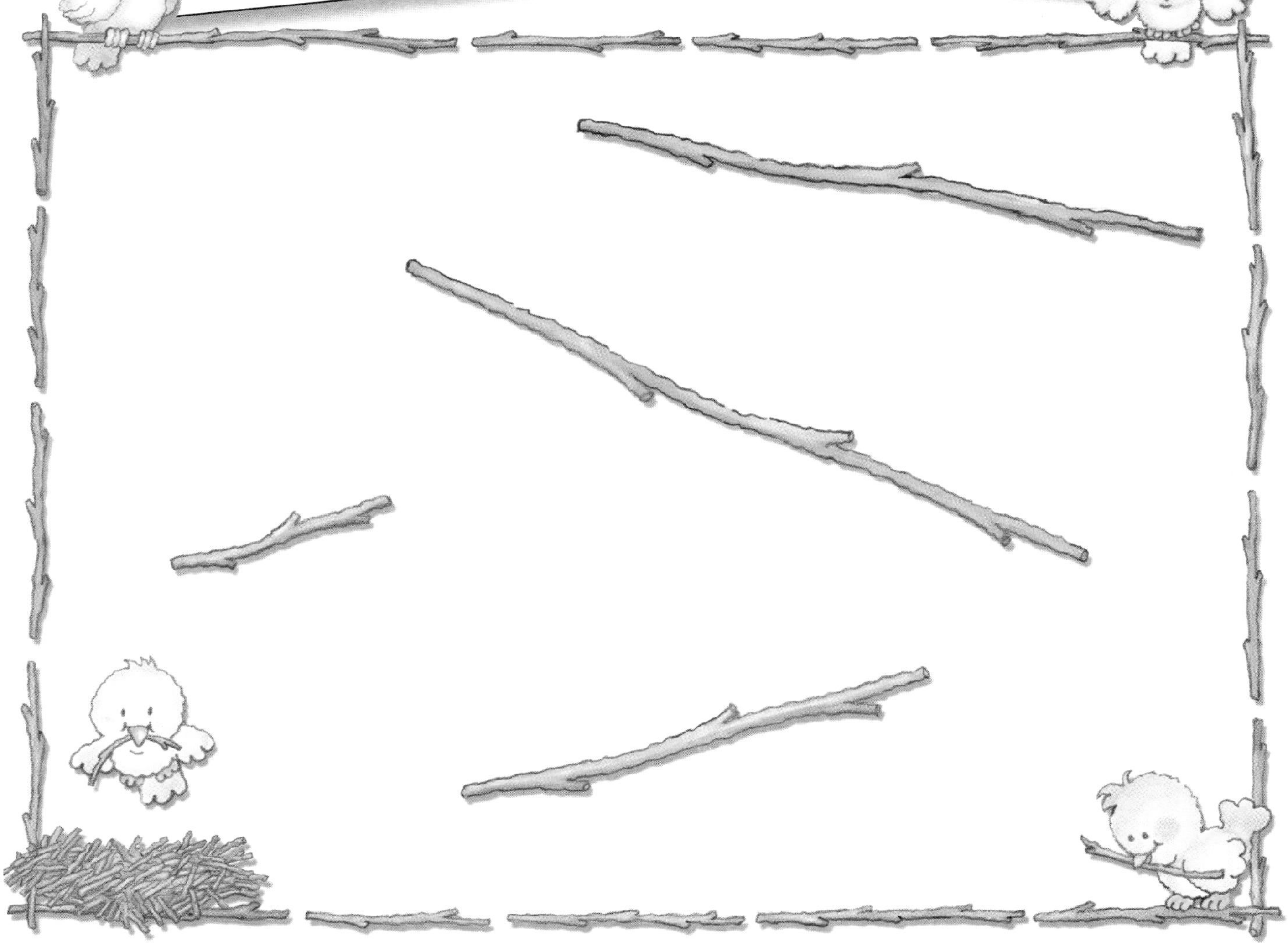

Look at each twig. Guess which twig is about five cubes long. Circle your guess. Use cubes to measure. Were you right?

Math at Home: Your child guessed about how long each twig was, and measured to check.
Activity: Use small objects, such as paper clips or beans. Ask your child to guess how many paper clips long a spoon is. Then have your child measure the spoon to check.

Problem Solving • Strategy

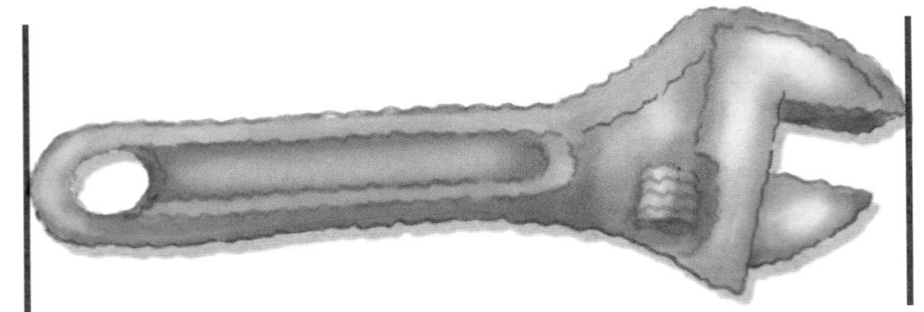

About ______  ______ 

About ______  ______ 

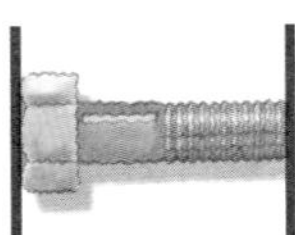

About ______  ______ 

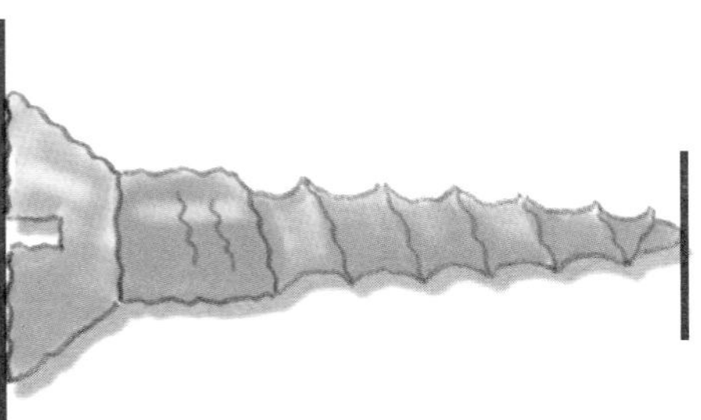

About ______  3 

Look at each object. Guess about how many cubes long. Write your guess.
Use cubes to measure each object. Write the number.

Name ______________________________

Color the containers that hold more blue
Color the containers that hold less red.

Math at Home: Your child explored capacity.
Activity: Show your child two containers and ask him or her to tell you which container holds more.

Circle each container that holds more.

Draw an X on each container that holds less.

Name ____________________

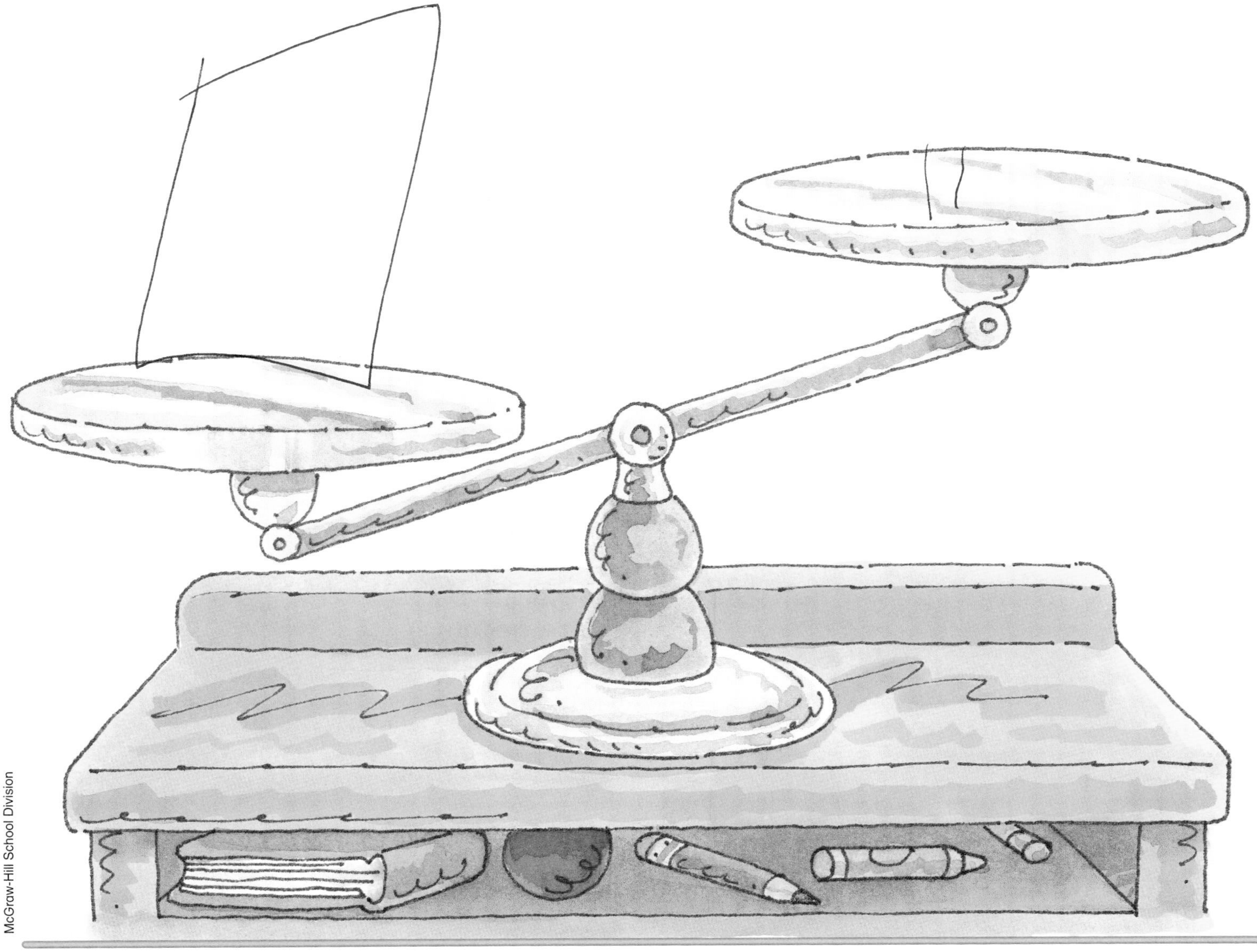

Find one heavy object and one light object in your classroom. Draw them on the balance.

Math at Home: Your child compared the weight of two objects.
Activity: Show your child two objects, such as a spoon and a can of beans. Have your child hold one object in each hand and tell which is heavier and which is lighter.

Look at the objects. Draw them on the balance.

Name ____________________

Math Words
hot
cold

Sna
San

hot

cold

Draw pictures to show what you might wear when it is hot and when it is cold.

Math at Home: Your child explored temperature using the words hot and cold.
Activity: Have your child find magazine pictures that show scenes that are hot and scenes that are cold.

Draw a circle around the pictures that show hot. Put an X on the pictures that show cold.

Name ____________________

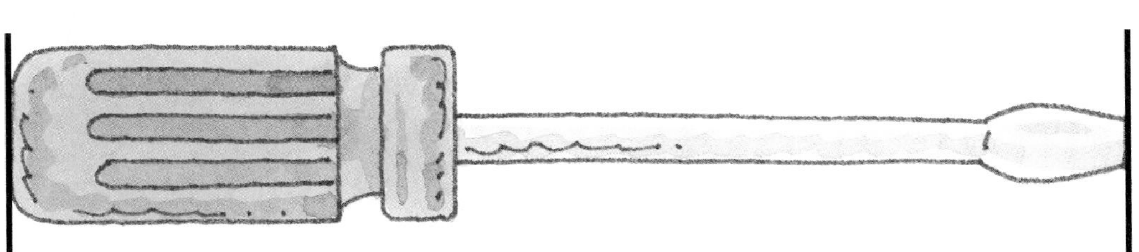

About ______ 

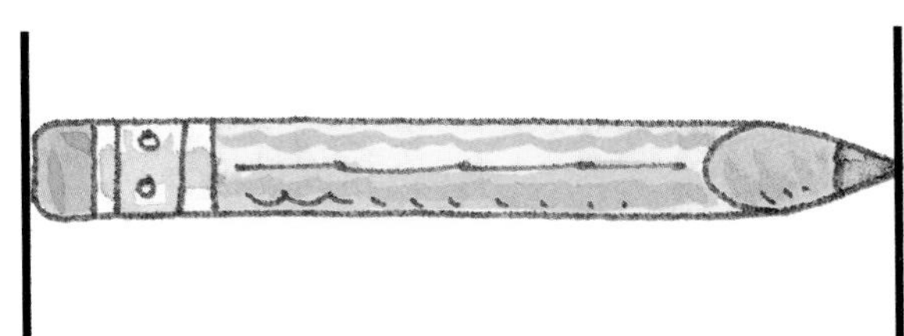

About ______ 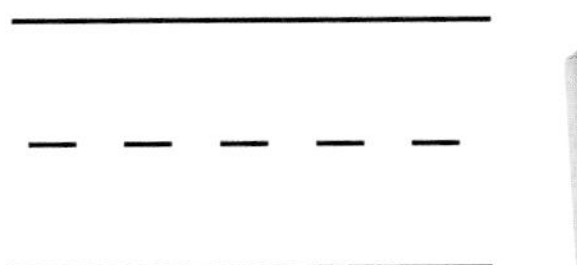 

Draw a circle around the longer object. Put an X on the shorter object.

Look at each object. Guess about how many cubes long. Write your guess. Use cubes to measure each object. Write the number.

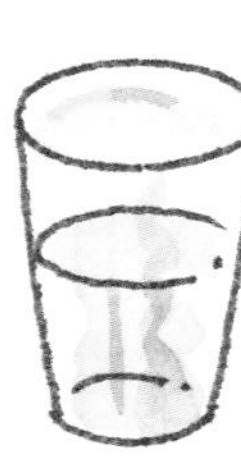

Draw a circle around the container that holds more. Put an X on the container that holds less. Look at the objects. Draw them on the balance. Draw a circle around the picture that shows something hot. Put an X on the picture that shows something cold.

CHAPTER 11

Time

theme
Going Places

Use the Data

What time is it?

What You Will Learn

In this chapter you will learn how to:

- Identify months and days.
- Tell time to the hour and half hour.
- Use logical reasoning to solve problems.

Dear Family,

In Chapter 11, I will learn about time and the calendar. Here are new vocabulary words and an activity that we can do together.

Math Words

calendar

month

day

minute hand

hour hand

Calendar Check

- Help your child name the current day, month, and year and identify the number of days in the month. Then mark special days on the calendar such as birthdays, holidays, or family outings.
- Look at the calendar together each day. Have your child mark off the current day. Ask your child to name the day of the week, and *today, yesterday,* and *tomorrow*. Ask you child how many more days until a special day.

use

calendar

crayons

Additional activities at www.mhschool.com/math

Name ______________________________

Say these words.

Math Words
calendar
day
month

April

Sunday	Monday	Tuesday	Wednesday	Thursday	Friday	Saturday
1	2	3	4	5	6	7
8	9	10	11	12	13	14
15	16	17	18	19	20	21
22	23	24	25	26	27	28
29	30					

Draw a line under the name of the month. Circle the name of each day of the week.
Color the first day of the month red. Color the last day of the month blue.

Math at Home: Your child identified the month and days on a calendar.
Activity: Show your child a calendar for the current month. Ask your child to show you the first and last days of the month.

May

Sunday	Monday	Tuesday	Wednesday	Thursday	Friday	Saturday
		1	2	3	4	5
6	7	8	9		11	12
13	14	15	16	17	18	
20		22	23	24	25	26
27	28	29		31		

Draw a picture of something you might do during this month. Write the missing dates.

Write the number of days in one week.

Write the number of days in this month.

Name ______________________________

Use Logical Reasoning

Look the pictures in each row. Write 1, 2, and 3 to show what happened first, next, and last.

Math at Home: Your child used logical reasoning to order events.
Activity: Ask your child how he or she decided which picture came first, next, and last.

Look at the pictures in each row. Write 1, 2, and 3 to show what happened first, next, and last.

Name ______________________________

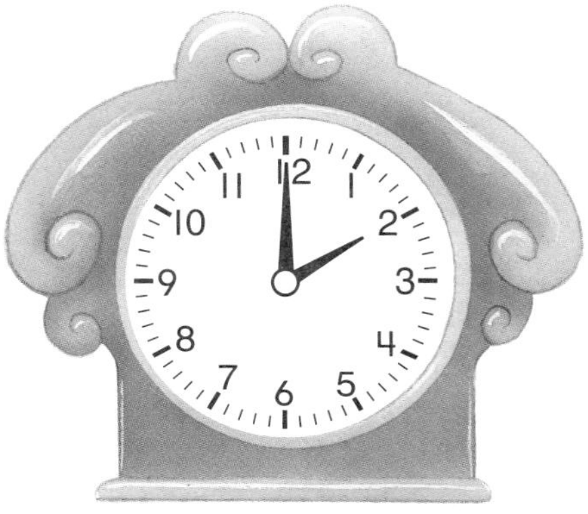

Math Words
hour hand
minute hand

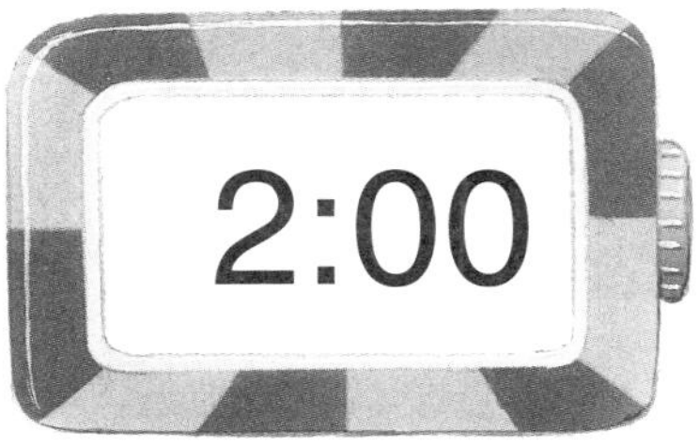

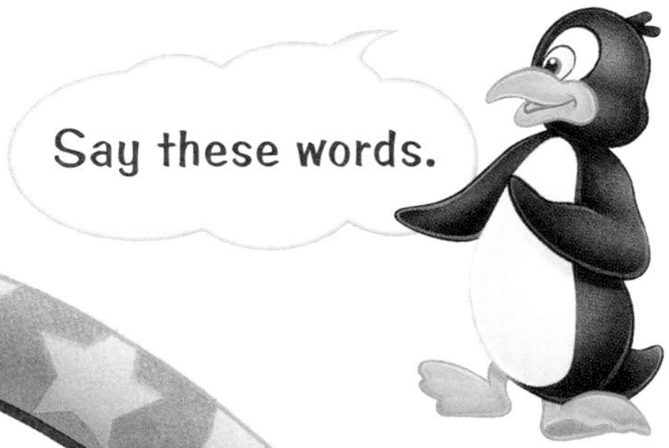

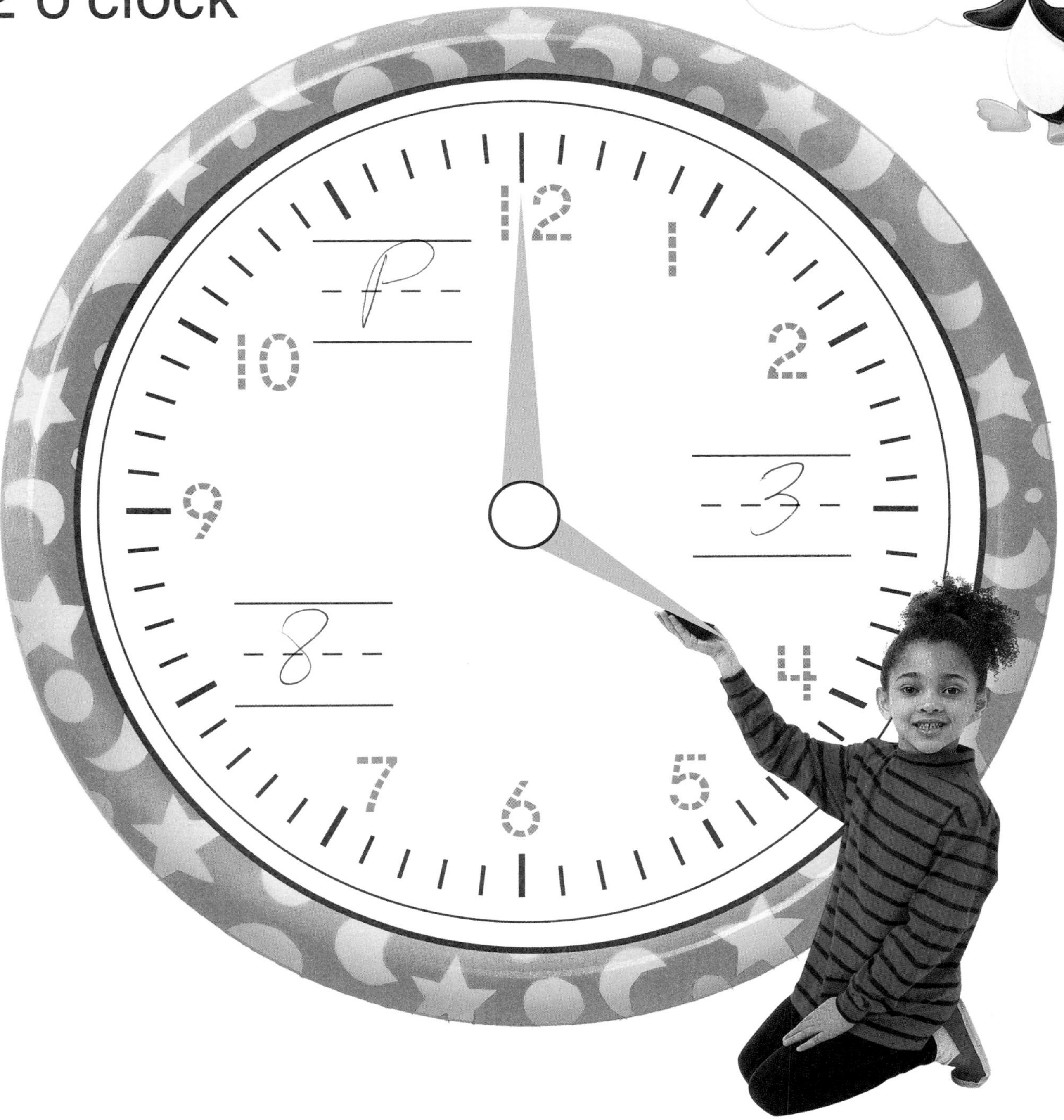

2:00

2 o'clock

2 o'clock

Say these words.

12 1 2 3 4 5 6 7 8 9 10

Write the numbers on the clock. Circle the number that the hour hand is pointing to. Draw an X through the number that the minute hand is pointing to.

Math at Home: Your child is learning to tell time.
Activity: Ask your child to find clocks in your home. Ask them what numbers the hands are pointing to.

Write the time shown on each clock. Circle the two clocks that show the same time.

Name ____________________

Tell Time to the Half Hour

two thirty

two thirty

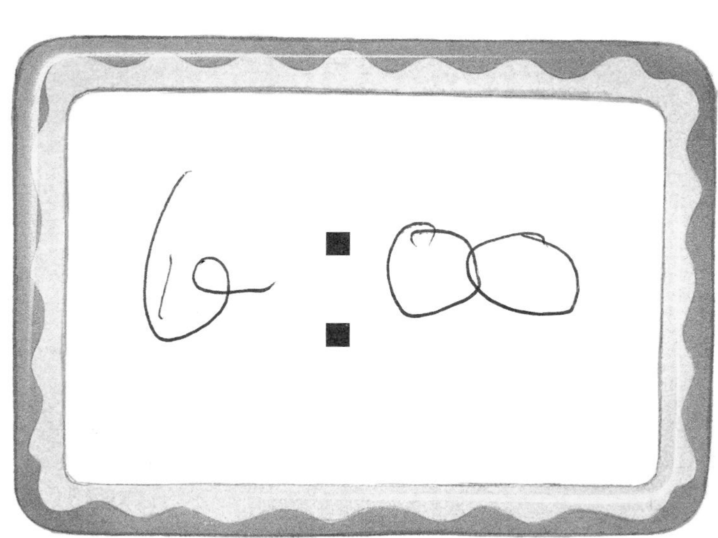

Draw hands or numbers on the clocks in each row to make them match.

Math at Home: Your child is learning how to tell time to the half hour.
Activity: Ask your child to show you which numbers on a digital clock show hours and which numbers show minutes.

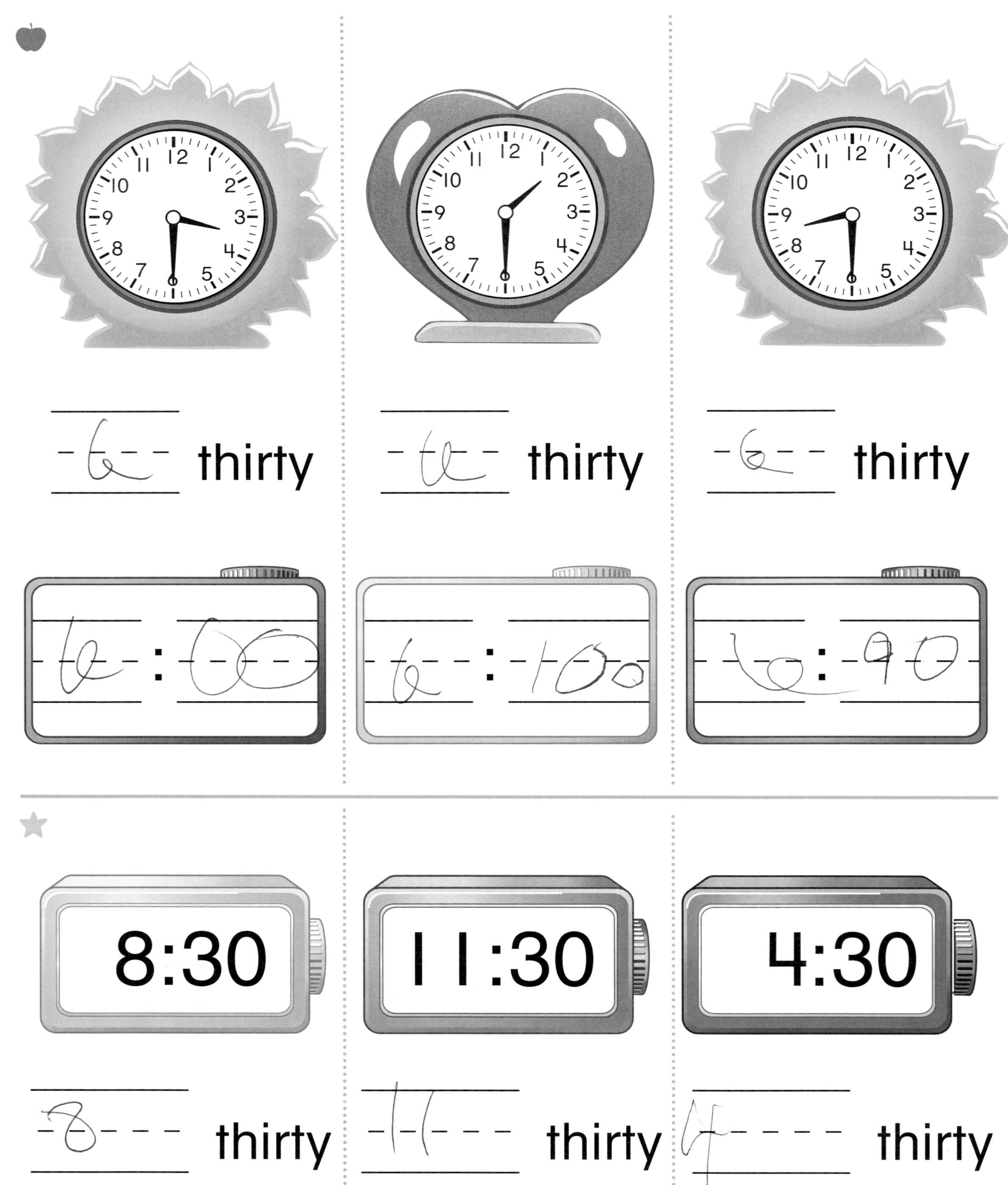

Write the time shown on each clock. Circle the two clocks that show the same time.

Name

June

Sunday	Monday	Tuesday	Wednesday	Thursday	Friday	Saturday
					1	2
3	4	5	6	7	8	9
10	11	12	13	14	15	16
17	18	19	20	21	22	23
24	25	26	27	28	29	30

Draw a line under the name of the month. Circle the name of each day of the week. Color the first day of the month red. Write the missing dates.

3 1 2

7:00

____ o'clock

5:30

____ thirty

9:30

____ thirty

- Look at each picture. Write 1, 2, or 3 to show what happened first, next, and last.
- Write the time shown on each clock. Circle the two clocks that show the same time.

CHAPTER 12

Addition Concepts

theme
In the Garden

Use the Data

How many vegetables are in the wheelbarrow?

What You Will Learn

In this chapter you will learn how to:

- Show the meaning of addition.
- Explore addition in horizontal and vertical forms.
- Add, sums to 10.
- Draw a picture to solve problems.

Dear Family,
In Chapter 12, I will learn about addition. Here are some new vocabulary words and an activity that we can do together.

Paper Plate Addition

- Place two paper plates side by side on a table. Give your child 5 small objects and have him or her place some of the objects on each plate.

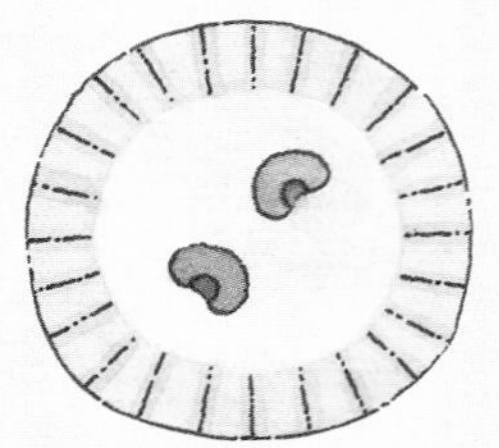

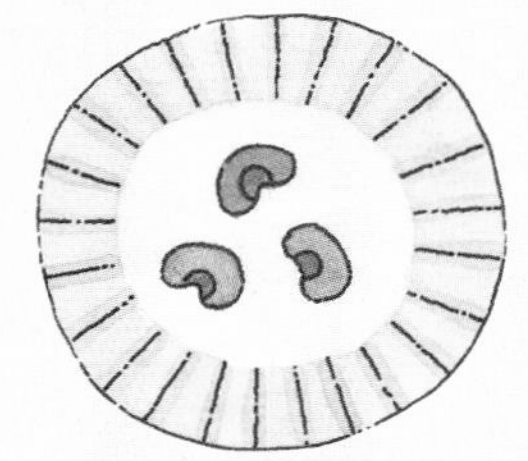

- Have your child tell you how many objects are on each plate, and how many there are altogether.
- Repeat the activity using different amounts of objects up to 10.

use

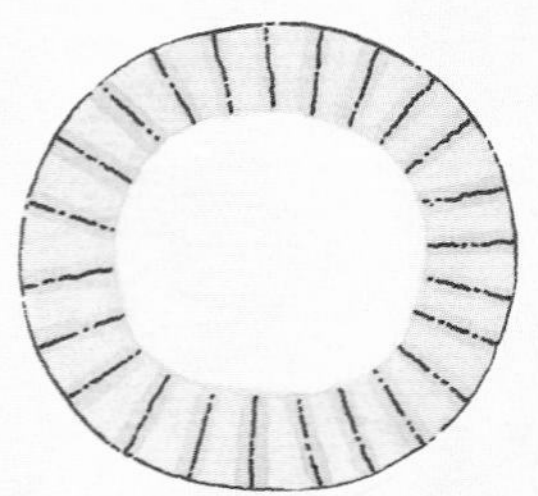

2 paper plates

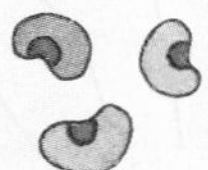

beans

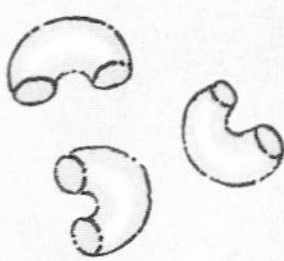

or pasta

Math Words

add

3 + 2 = 5

$$\begin{array}{r} 3 \\ +\,2 \\ \hline 5 \end{array}$$

plus +
3 **plus** 2 equals 5.

equals =
3 plus 2 **equals** 5.

sum
3 + 2 = 5 ← sum

The sum tells how many there are in all.

Additional activities at www.mhschool.com/math

Name____

3 and 2 is 5

2 and 1 is 3

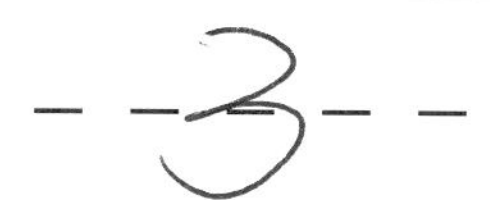

2 and 2 is 4

Tell how many are in each group. Tell how many in all. Write the numbers.

Math at Home: Your child is learning how to join groups.
Activity: Make 2 groups from 5 small objects. Ask your child to tell how many there are in each group and then how many there are in all.

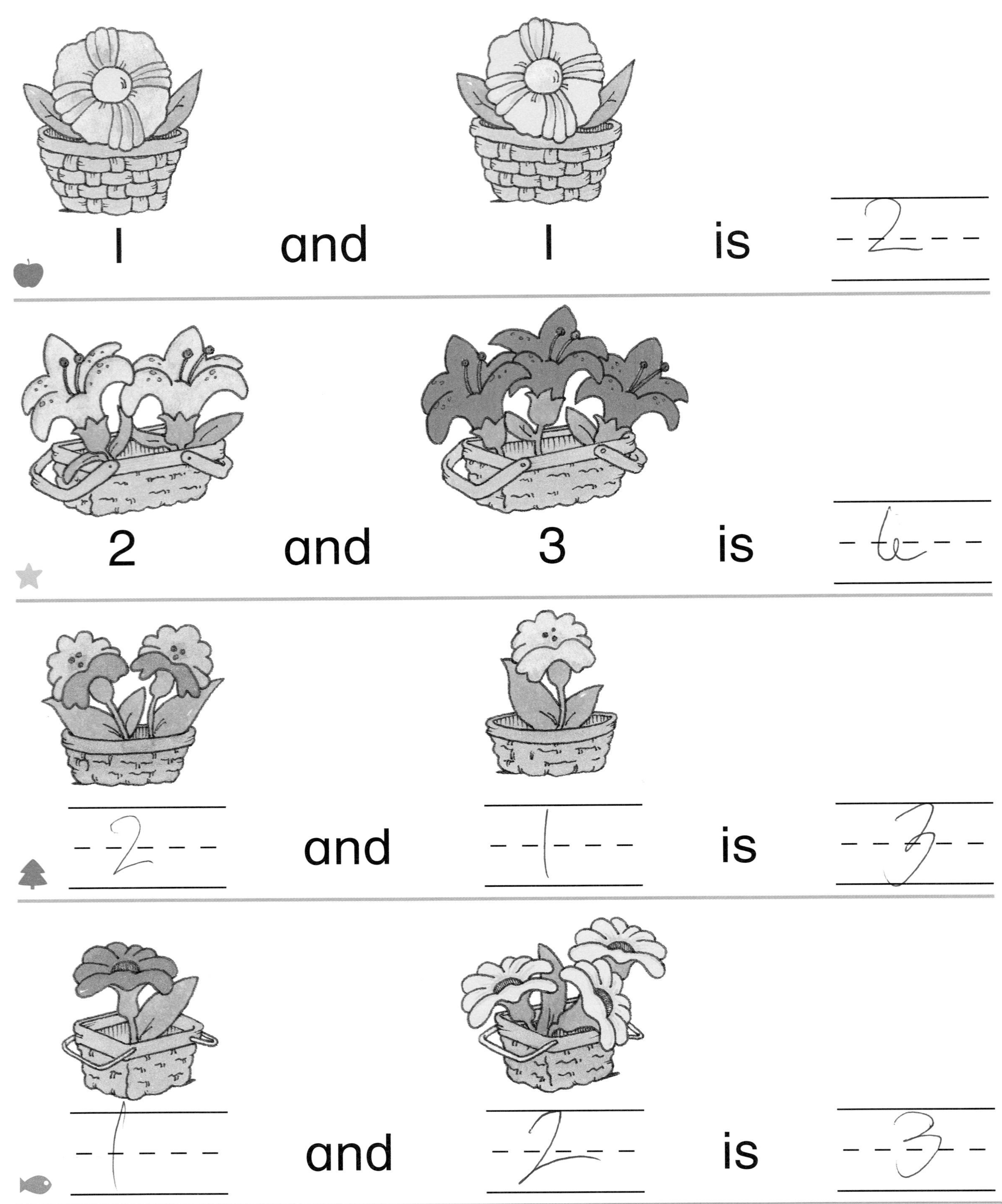

Tell how many are in each group. Tell how many in all. Write the numbers.

Name ______________________

3 and 1 is 4

4 and 1 is 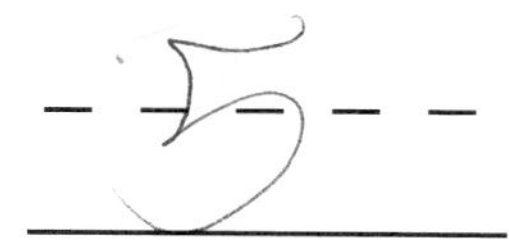

 and 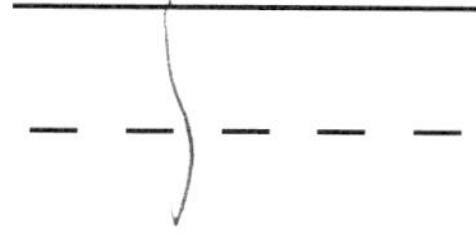 is 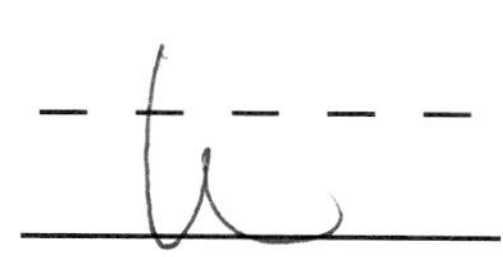

Tell a story about each picture. Count on to find how many there are in all. Write the numbers.

Math at Home: Your child added by counting on 1.
Activity: Show your child up to 9 small objects and then add 1 more. Ask your child to tell how many there are in all.

6 and 1 is ______

1 and 1 is ______

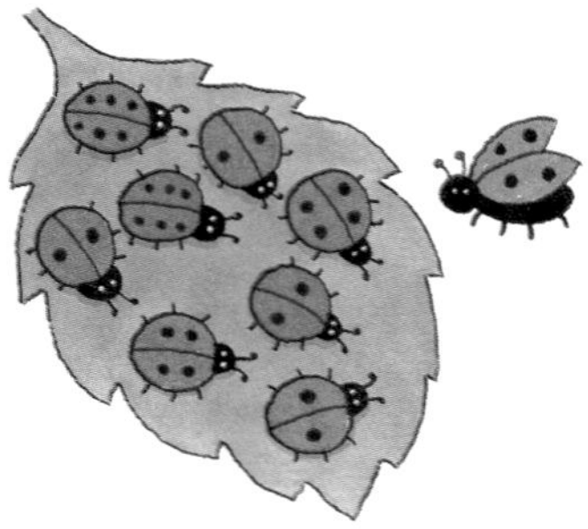

8 and 1 is ______

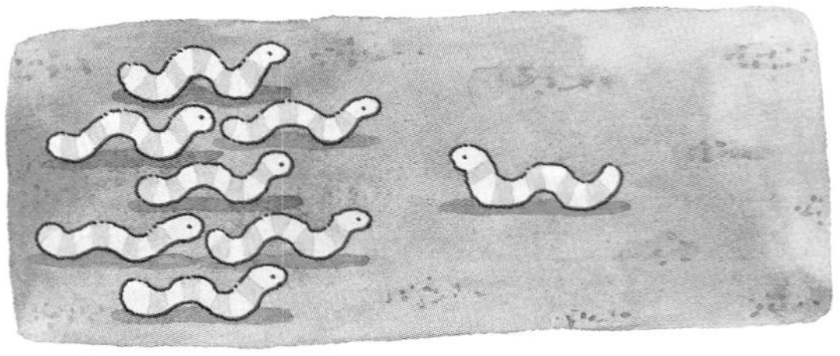

7 and 1 is ______

______ and ______ is ______

______ and ______ is ______

Tell a story about each picture. Count on to find how many there are in all. Write the numbers.

Name ________________________________

4 and 2 is 6

7 and 1 is ______

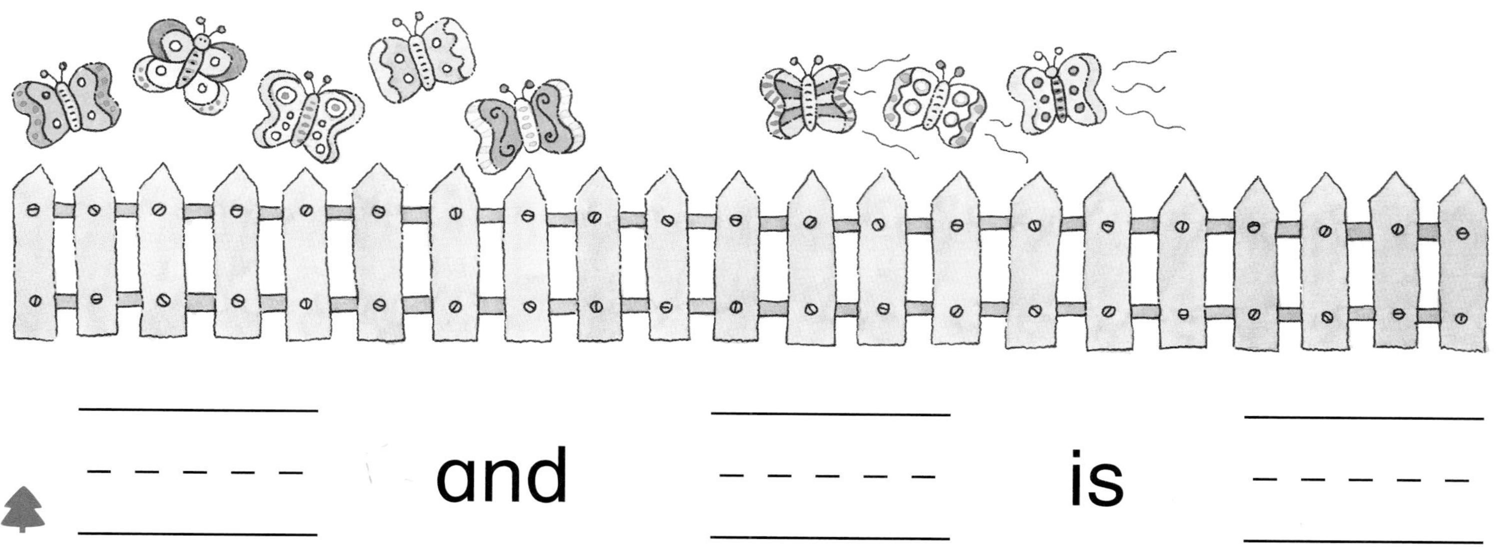

______ and ______ is ______

Tell a story about each picture. Count on to find how many there are in all. Write the numbers.

Math at Home: Your child added by counting on 1, 2, or 3.
Activity: Show your child a group of up to 7 small objects. Then show another group of 1, 2, or 3 objects. Ask your child to tell how many there are in all.

5 and 1 is ______

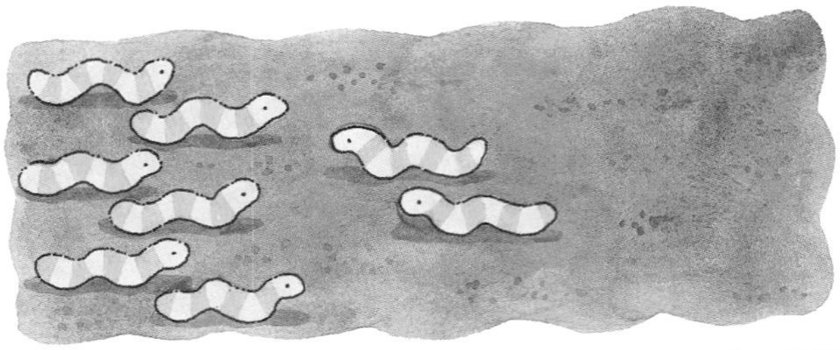

6 and 2 is ______

4 and 3 is ______

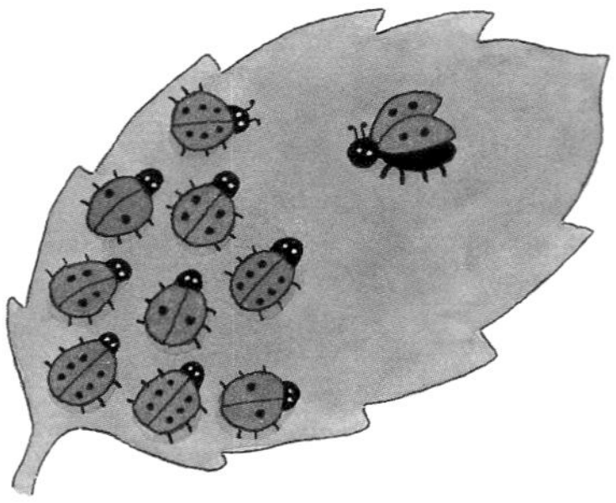

9 and 1 is ______

______ and ______ is ______

______ and ______ is ______

Tell a story about each picture. Count on to find how many there are in all. Write the numbers.

Name ______________________________

Math Words
add
plus
equals
sum

Say these words.

2 + 1 = 3

3 + 1 = ____

3 + 2 = ____

Tell a story about each picture. Add to find how many there are in all. Write each sum.

Math at Home: Your child learned addition and the + and = signs.
Activity: Ask your child to tell you what each sign means.

4 + 1 = ______

1 + 2 = ______

2 + 3 = ______

1 + 1 = ______

2 + 2 = ______

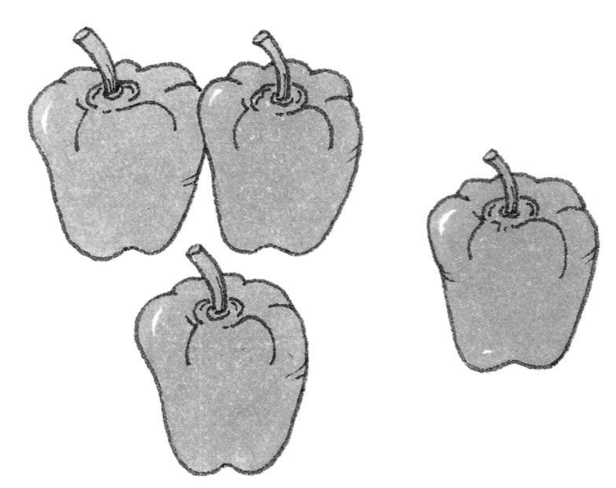

3 + 1 = ______

Tell a story about each picture. Add to find how many in all. Write each sum.

Name ____________________

Draw a Picture

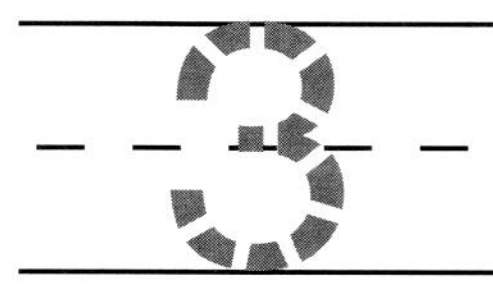  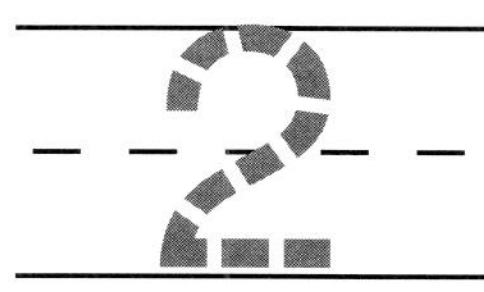 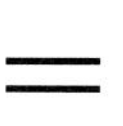 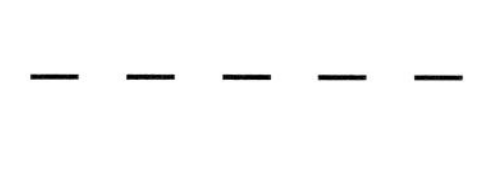

Listen to the story. Draw a picture to solve the problem. Write the addition sentence.

Math at Home: Your child drew pictures to solve problems.
Activity: Make up a story problem such as: There are 2 red balls. There are 2 green balls. How many balls are there in all? Have your child draw a picture to solve the problem and write the addition sentence.

Listen to each story. Draw a picture to solve the problem. Write the addition sentence.

Name ______________________________

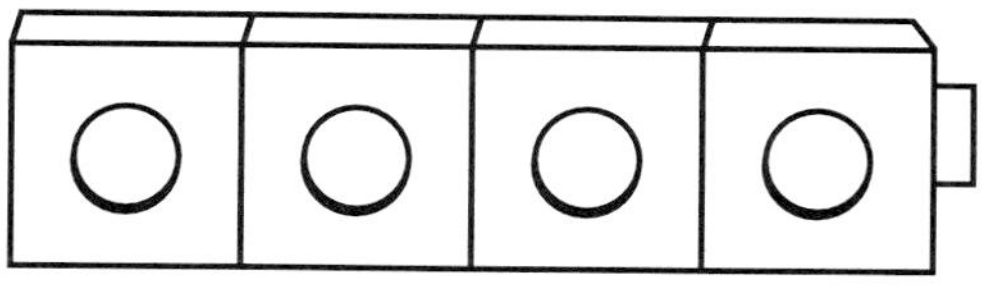

4

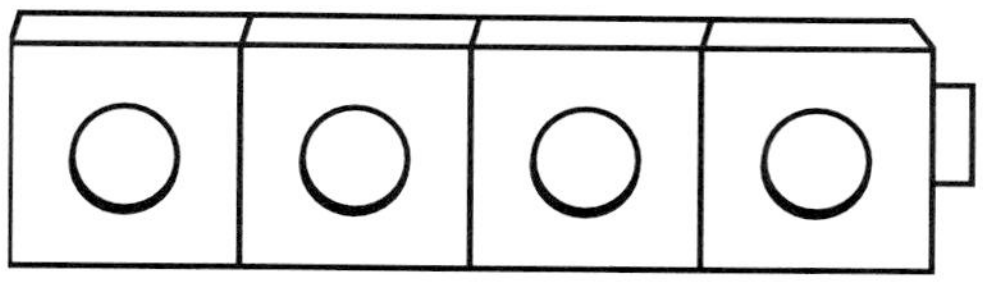

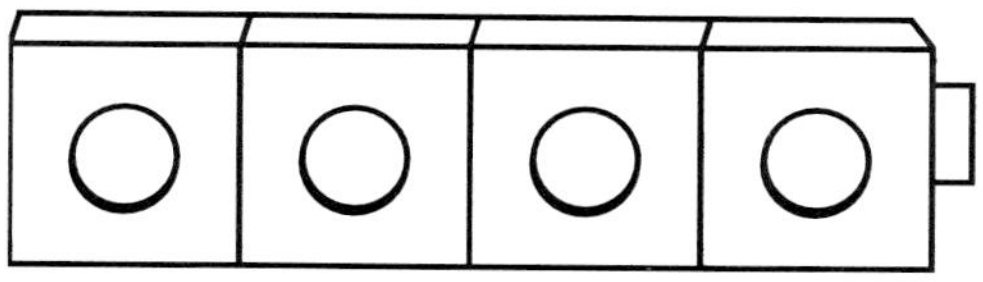

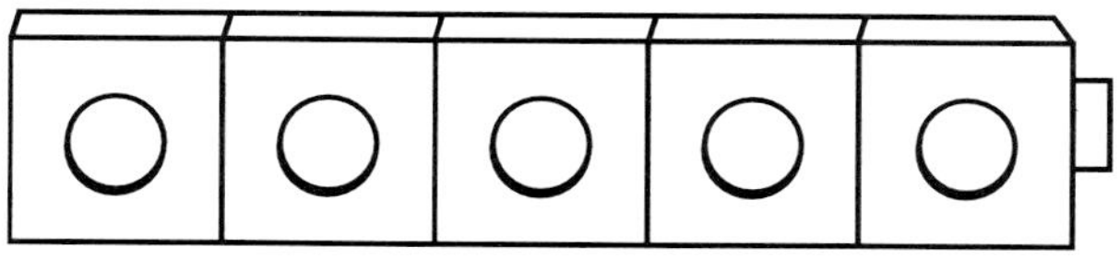

5

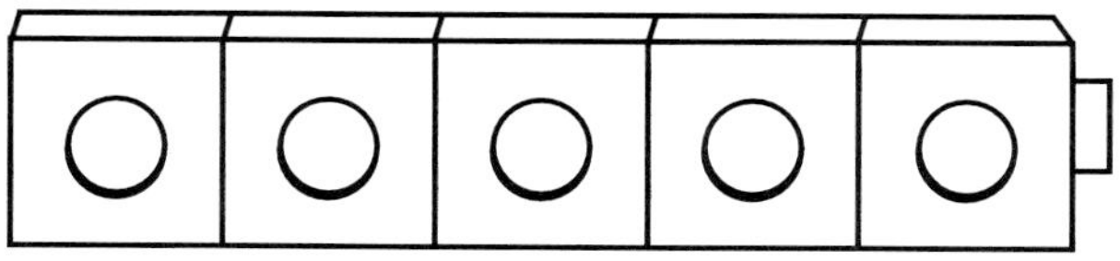

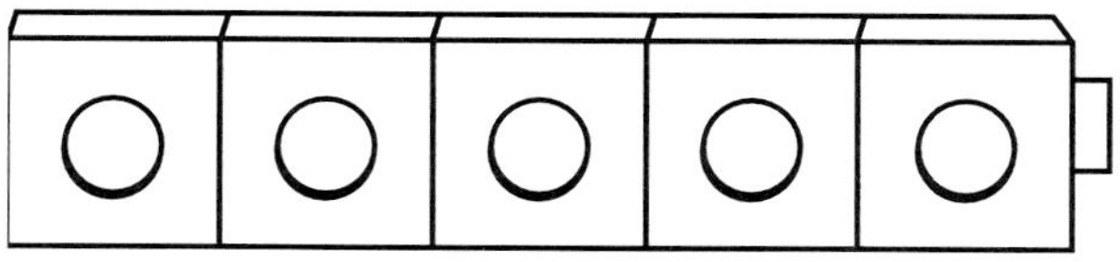

Use 2 different colors of cubes to make 4 and 5. Color.

Math at Home: Your child used cubes to show different ways to make 4 and 5.
Activity: Ask your child to use two different kinds of pasta pieces to show ways to make 4 and 5.

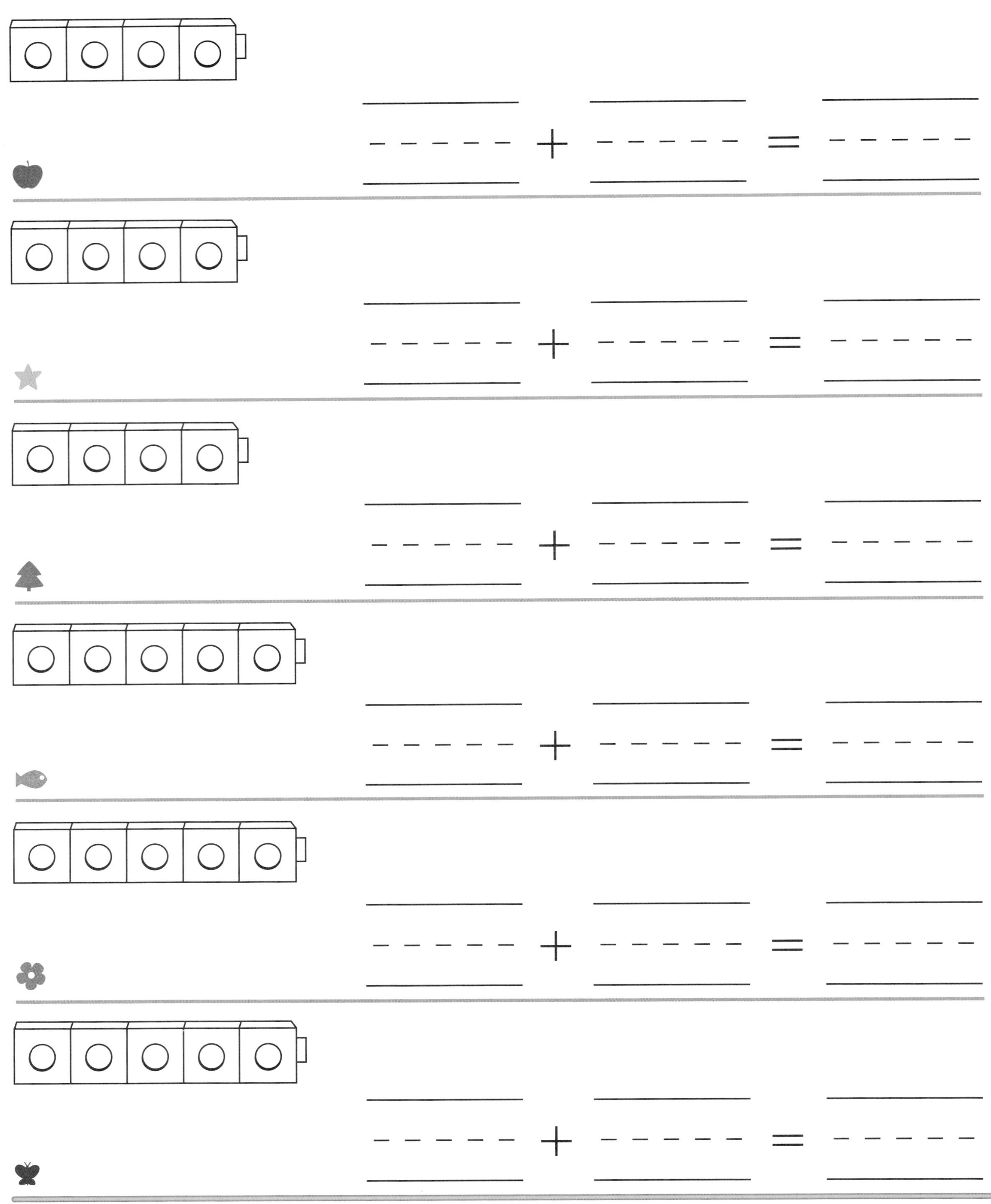

Use 2 different colors of cubes to make 4 and 5. Color. Write the addition sentences.

Name ______________________________

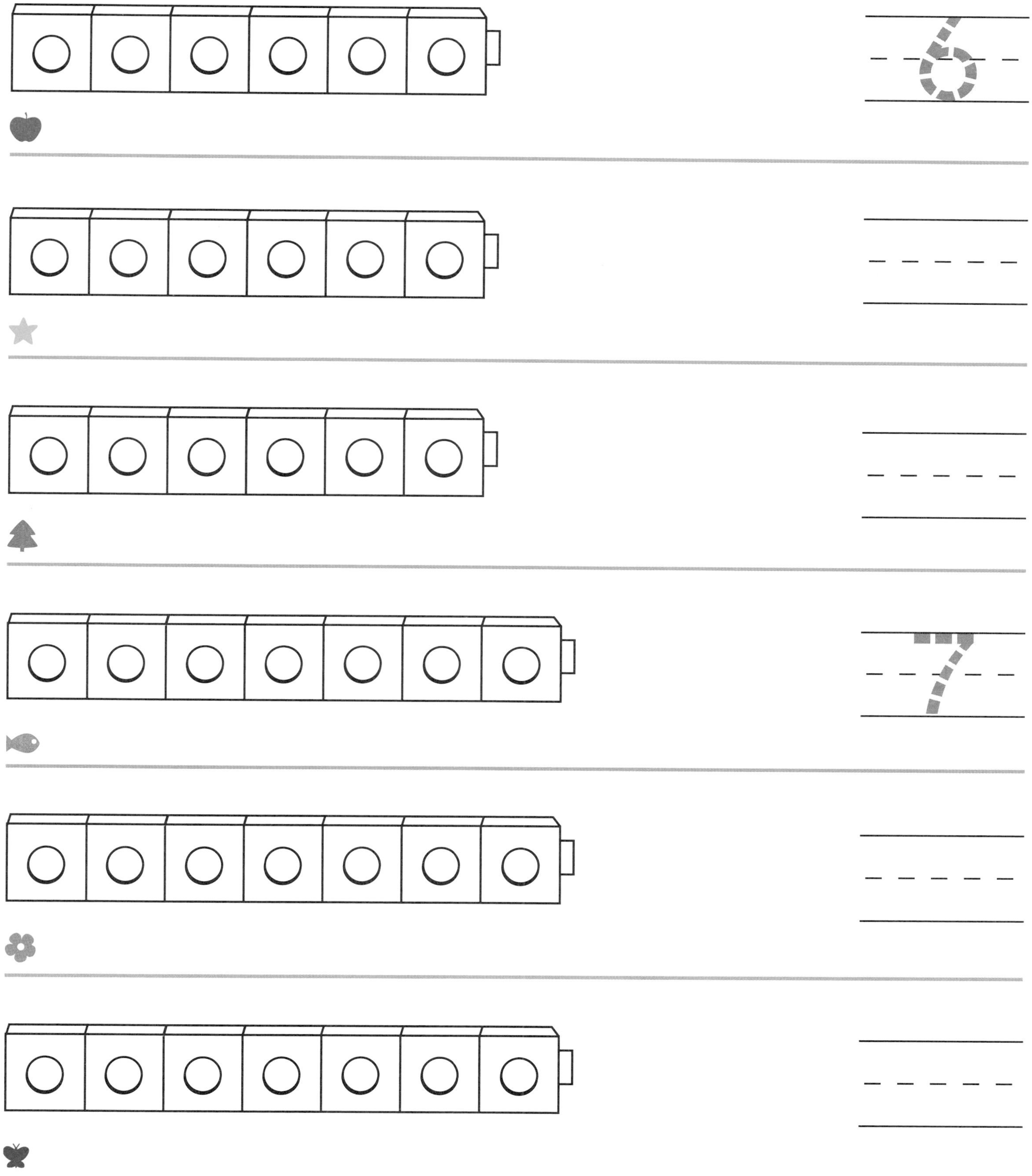

Use 2 different colors of cubes to make 6 and 7. Color.

Math at Home: Your child used cubes to show different ways to make 6 and 7.
Activity: Ask your child to use two different kinds of pasta pieces to show ways to make 6 and 7.

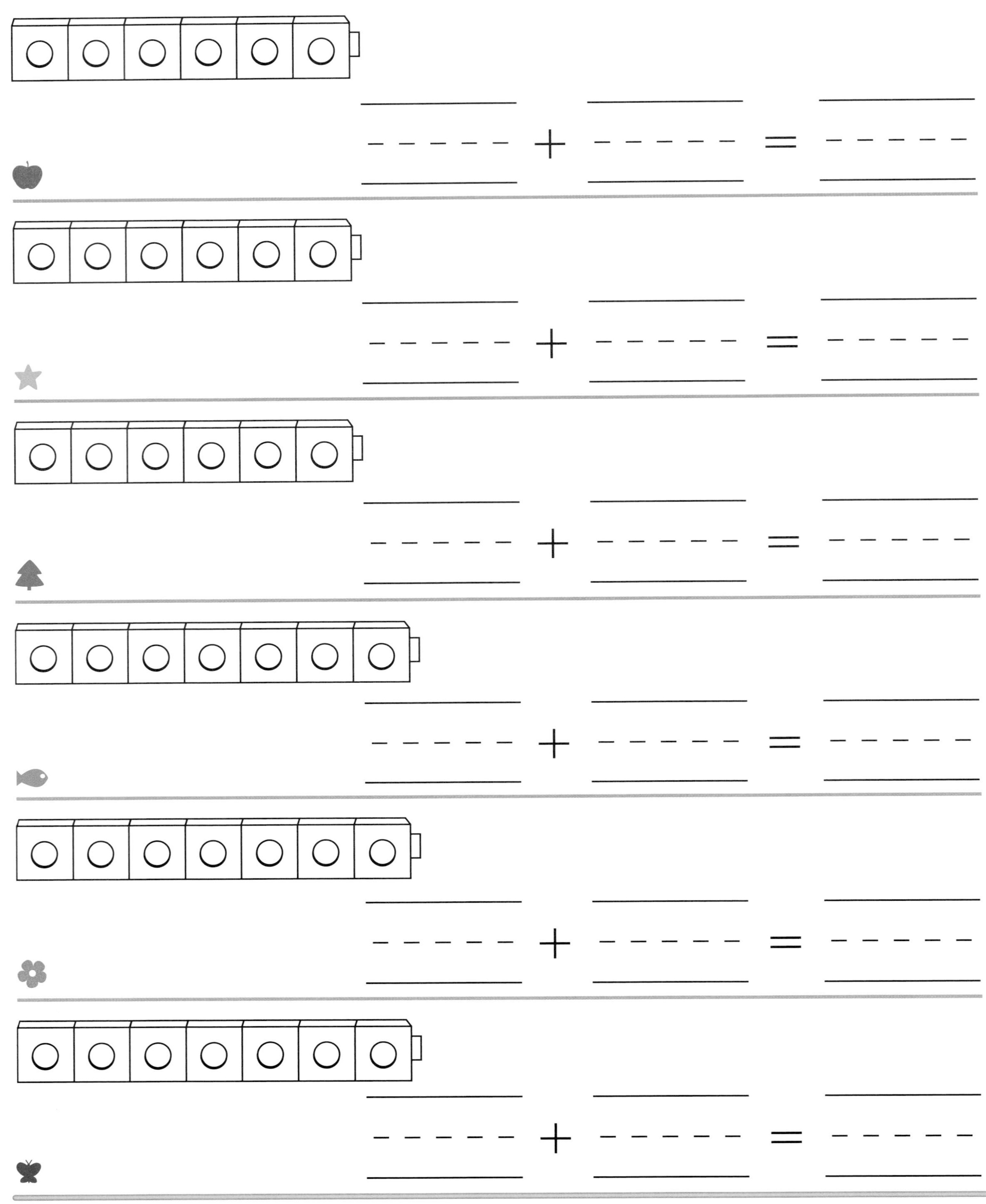

Use 2 different colors of cubes to make 6 and 7. Color. Write the addition sentences.

Name ____________________

8

9

Use 2 different colors of cubes to make 8 and 9. Color.

Math at Home: Your child used cubes to show different ways to make 8 and 9.
Activity: Ask your child to use two different kinds of pasta pieces to show ways to make 8 and 9.

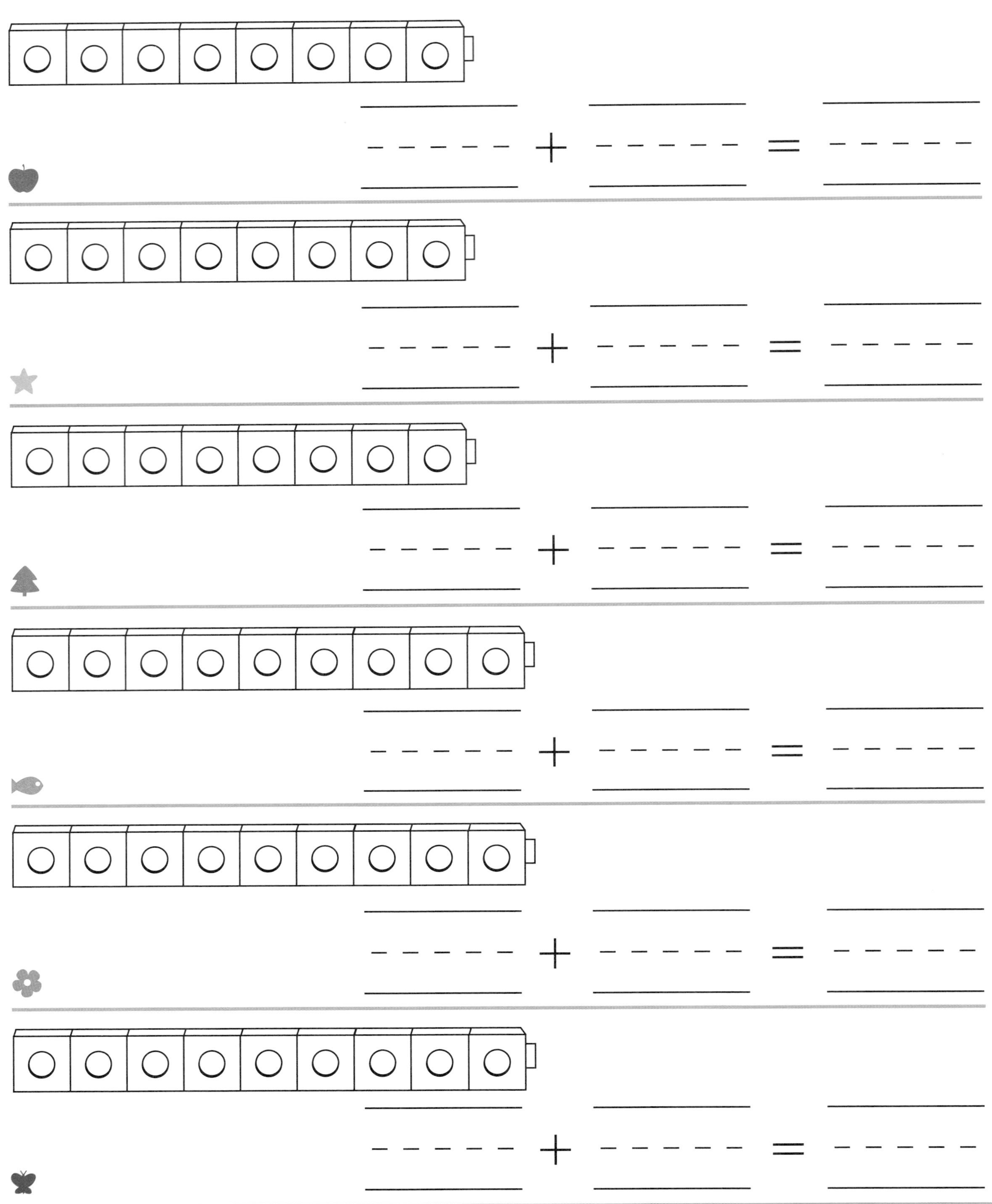

Use 2 different colors of cubes to make 8 and 9. Color. Write the addition sentences.

Name ______________________________

Ways to Make 10

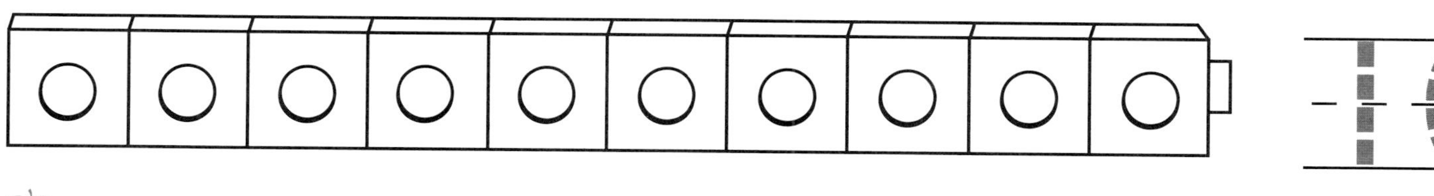

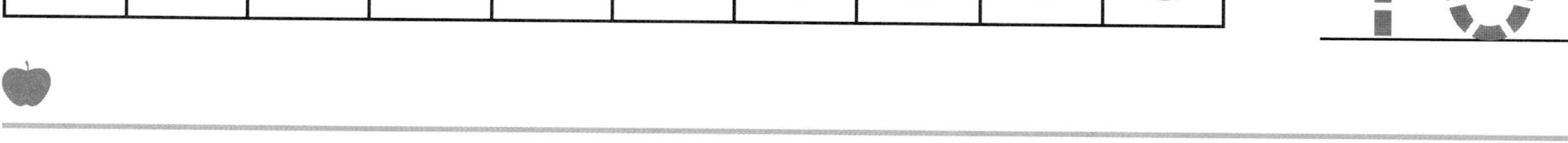

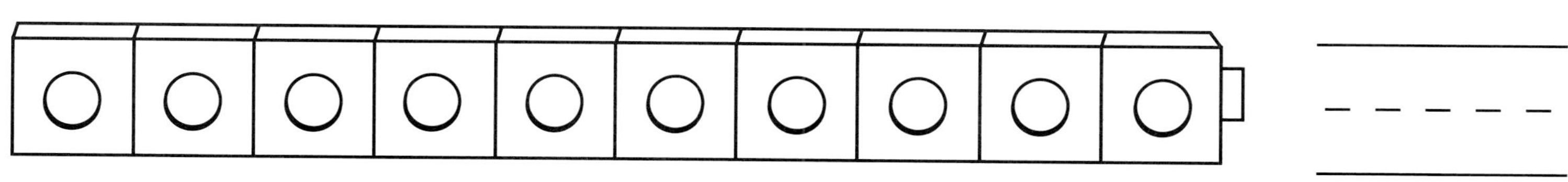

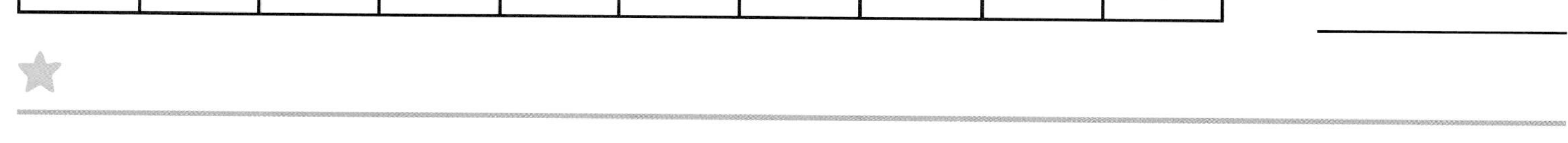

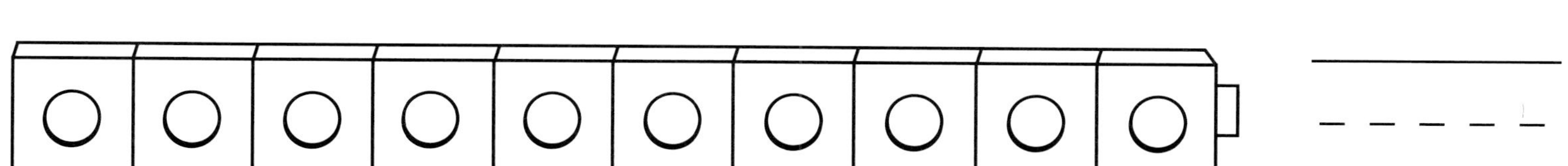

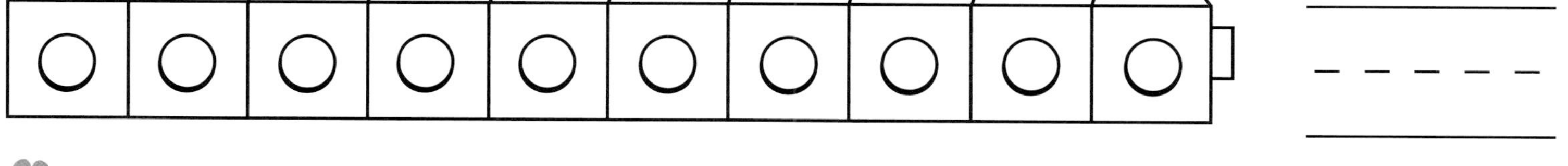

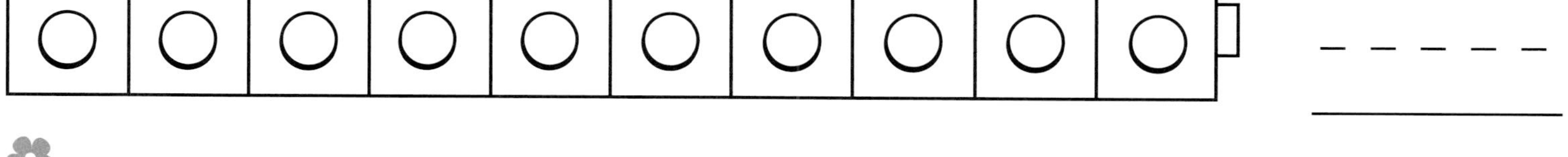

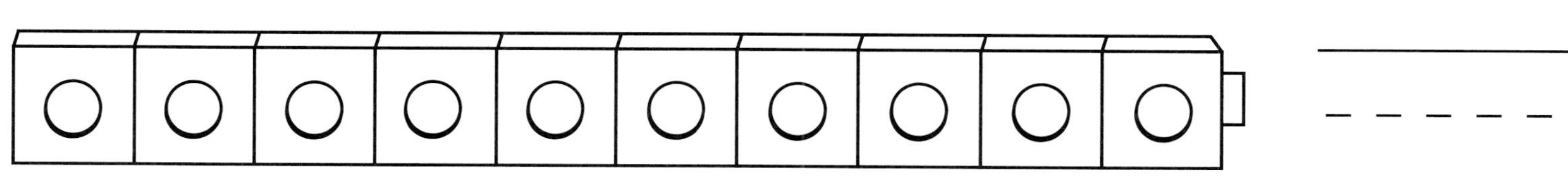

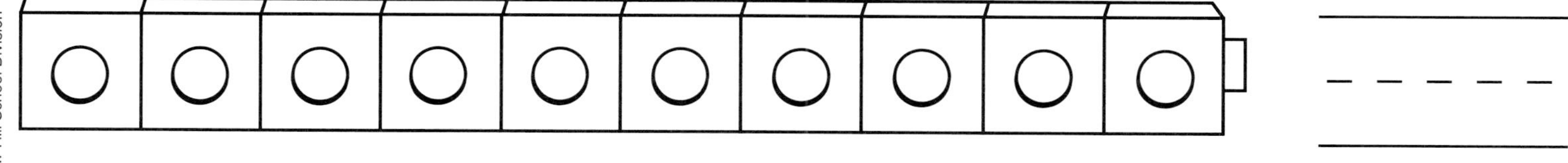

Use 2 different colors of cubes to make 10. Color.

Math at Home: Your child used cubes to show different ways to make 10.
Activity: Ask your child to use two different kinds of pasta pieces to show ways to make 10.

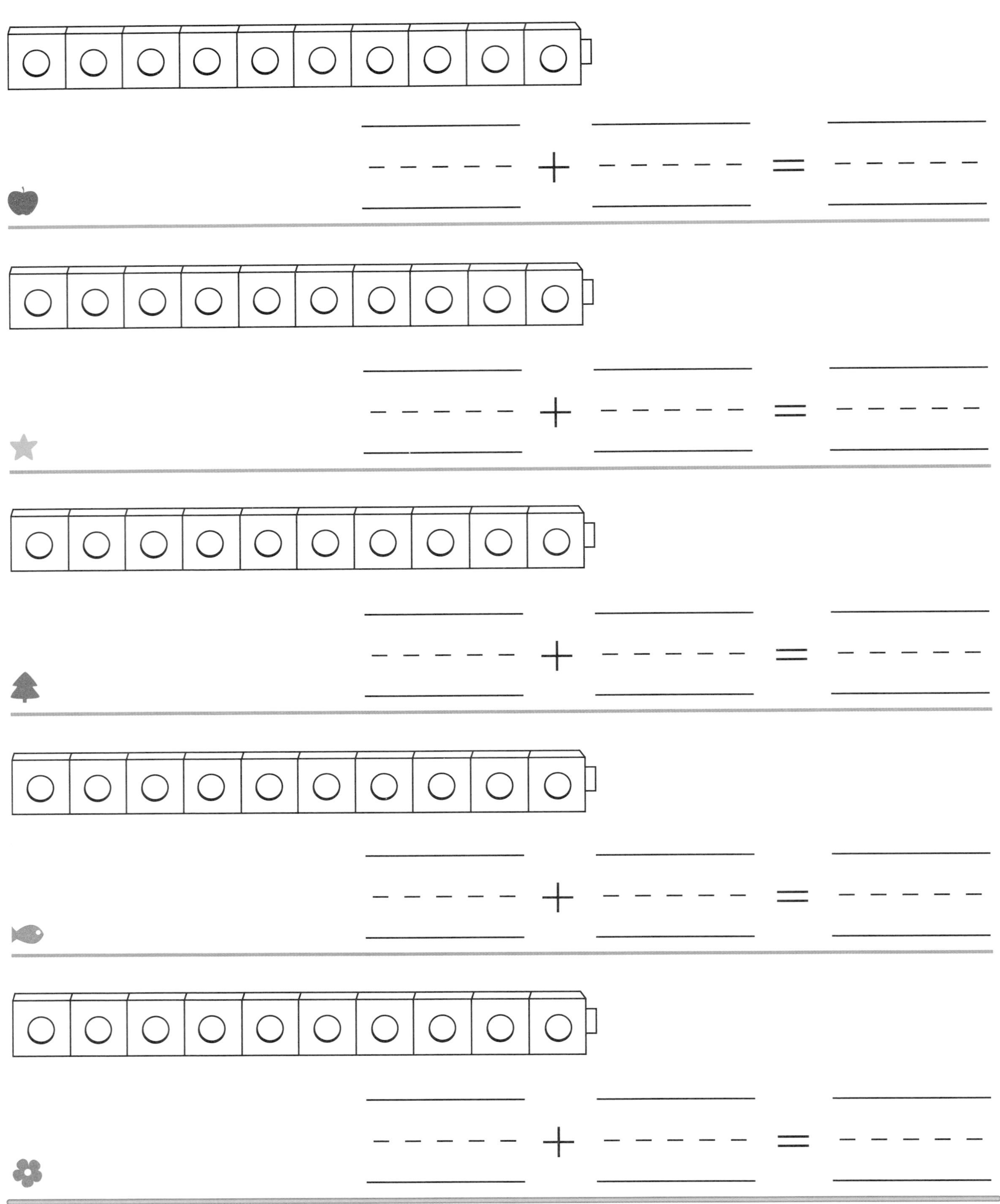

Use 2 different colors of cubes to make 10. Color. Write the addition sentences.

Name ______________________________

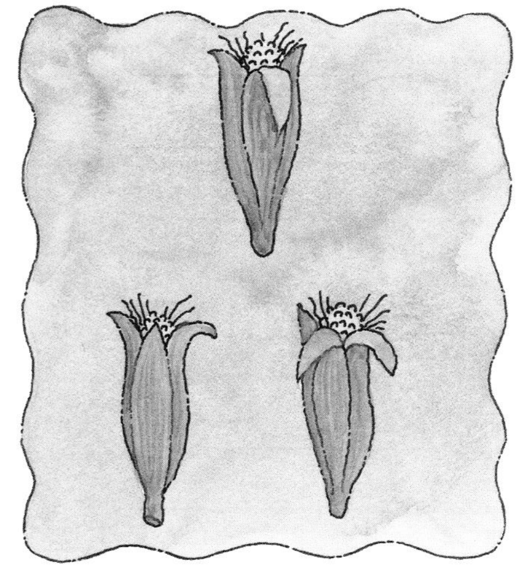

$$\begin{array}{r} 1 \\ +\,2 \\ \hline \end{array}$$

______

$$\begin{array}{r} 3 \\ +\,1 \\ \hline \end{array}$$

______

$$\begin{array}{r} 2 \\ +\,2 \\ \hline \end{array}$$

______

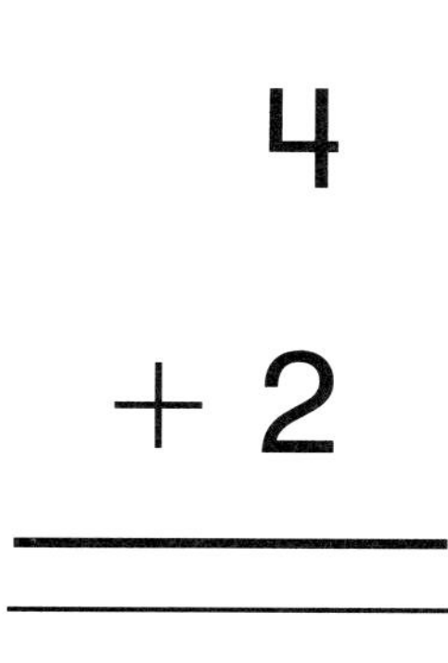

$$\begin{array}{r} 4 \\ +\,2 \\ \hline \end{array}$$

______

Look at each picture. Add.

Math at Home: Your child is learning vertical addition.
Activity: Write a vertical addition problem with a sum to ten and ask your child to tell you the answer.

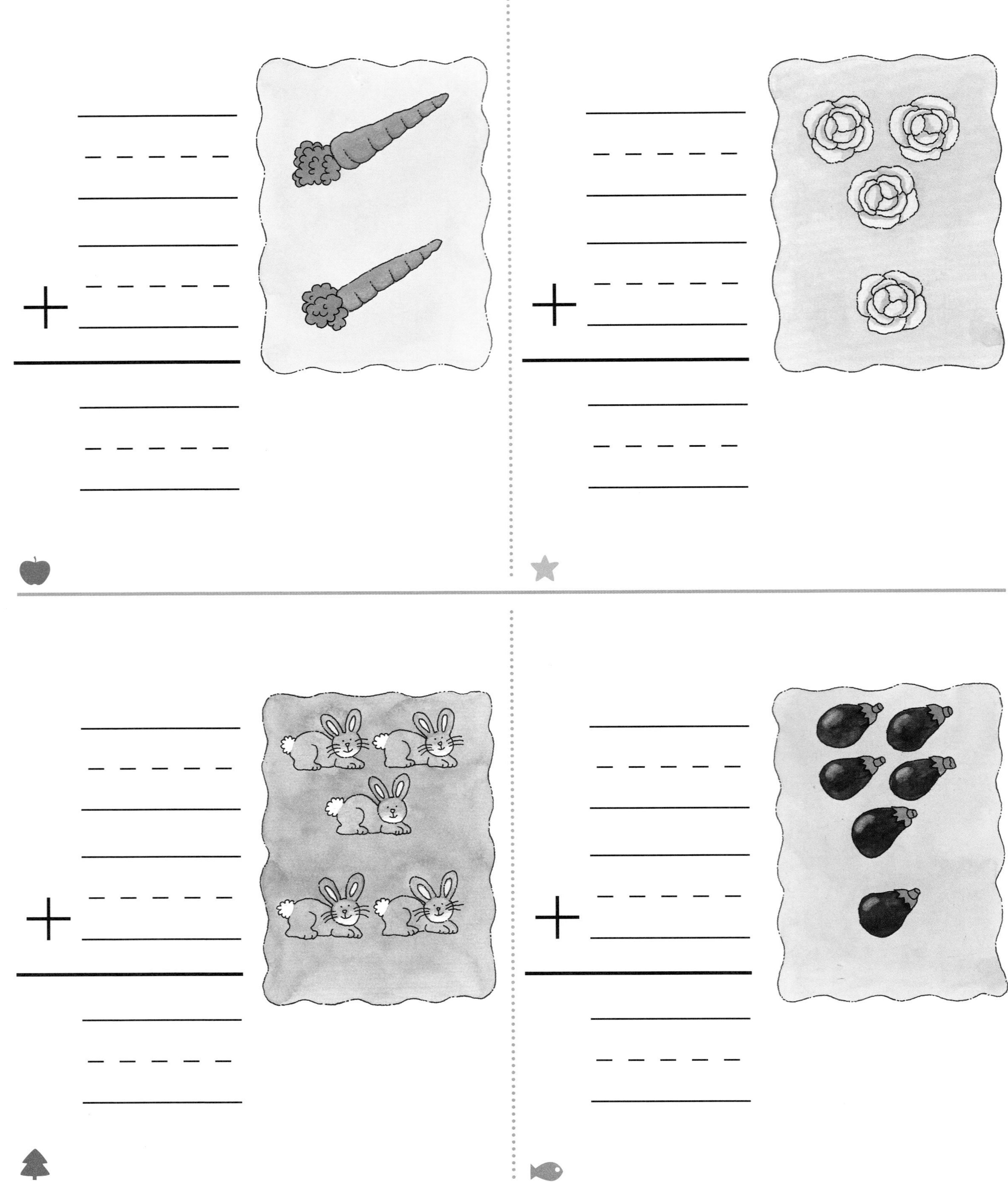

Look at each picture. Write the addition.

Name ______________________

$2 + 3 = 5$

$$\begin{array}{r} 2 \\ +\ 3 \\ \hline 5 \end{array}$$

$3 + 1 =$ ______

$$\begin{array}{r} 3 \\ +\ 1 \\ \hline \end{array}$$

$2 + 2 =$ ______

$$\begin{array}{r} 2 \\ +\ 2 \\ \hline \end{array}$$

$1 + 4 =$ ______

$$\begin{array}{r} 1 \\ +\ 4 \\ \hline \end{array}$$

$5 + 1 =$ ______

$$\begin{array}{r} 5 \\ +\ 1 \\ \hline \end{array}$$

Add. You can use cubes.

Math at Home: Your child practiced adding numbers horizontally and vertically.
Activity: Point to a problem on this page and ask your child to tell you how to find the sum.

$$\begin{array}{r} 4 \\ +\ 2 \\ \hline \end{array} \qquad \begin{array}{r} 1 \\ +\ 5 \\ \hline \end{array} \qquad \begin{array}{r} 3 \\ +\ 3 \\ \hline \end{array} \qquad \begin{array}{r} 2 \\ +\ 3 \\ \hline \end{array}$$

$$\begin{array}{r} 1 \\ +\ 3 \\ \hline \end{array} \qquad \begin{array}{r} 3 \\ +\ 4 \\ \hline \end{array} \qquad \begin{array}{r} 2 \\ +\ 1 \\ \hline \end{array} \qquad \begin{array}{r} 5 \\ +\ 1 \\ \hline \end{array}$$

$$\begin{array}{r} 3 \\ +\ 2 \\ \hline \end{array} \qquad \begin{array}{r} 2 \\ +\ 6 \\ \hline \end{array} \qquad \begin{array}{r} 4 \\ +\ 1 \\ \hline \end{array} \qquad \begin{array}{r} 1 \\ +\ 1 \\ \hline \end{array}$$

$$\begin{array}{r} 1 \\ +\ 6 \\ \hline \end{array} \qquad \begin{array}{r} 4 \\ +\ 5 \\ \hline \end{array} \qquad \begin{array}{r} 2 \\ +\ 4 \\ \hline \end{array} \qquad \begin{array}{r} 1 \\ +\ 2 \\ \hline \end{array}$$

Add. You can use cubes.

Name ____________________

4¢ + 2¢ = ______ ¢

$$\begin{array}{r} 4¢ \\ +\ 2¢ \\ \hline 6¢ \end{array}$$

3¢ + 2¢ = ______ ¢

$$\begin{array}{r} 3¢ \\ +\ 2¢ \\ \hline \_\_\_¢ \end{array}$$

3¢ + 1¢ = ______ ¢

$$\begin{array}{r} 3¢ \\ +\ 1¢ \\ \hline \_\_\_¢ \end{array}$$

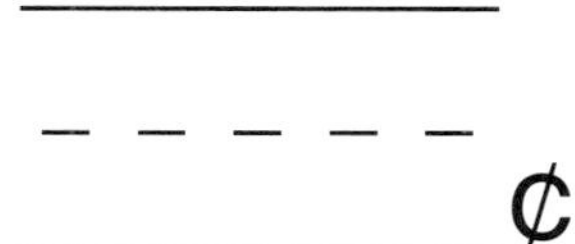

2¢ + 3¢ = ______ ¢

$$\begin{array}{r} 2¢ \\ +\ 3¢ \\ \hline \_\_\_¢ \end{array}$$

Add the money amounts. Write the sum.

Math at Home: Your child added pennies.
Activity: Show your child two groups of pennies that total 10¢ or less. Ask your child to tell you how many pennies are in each group and then add the pennies to find the total.

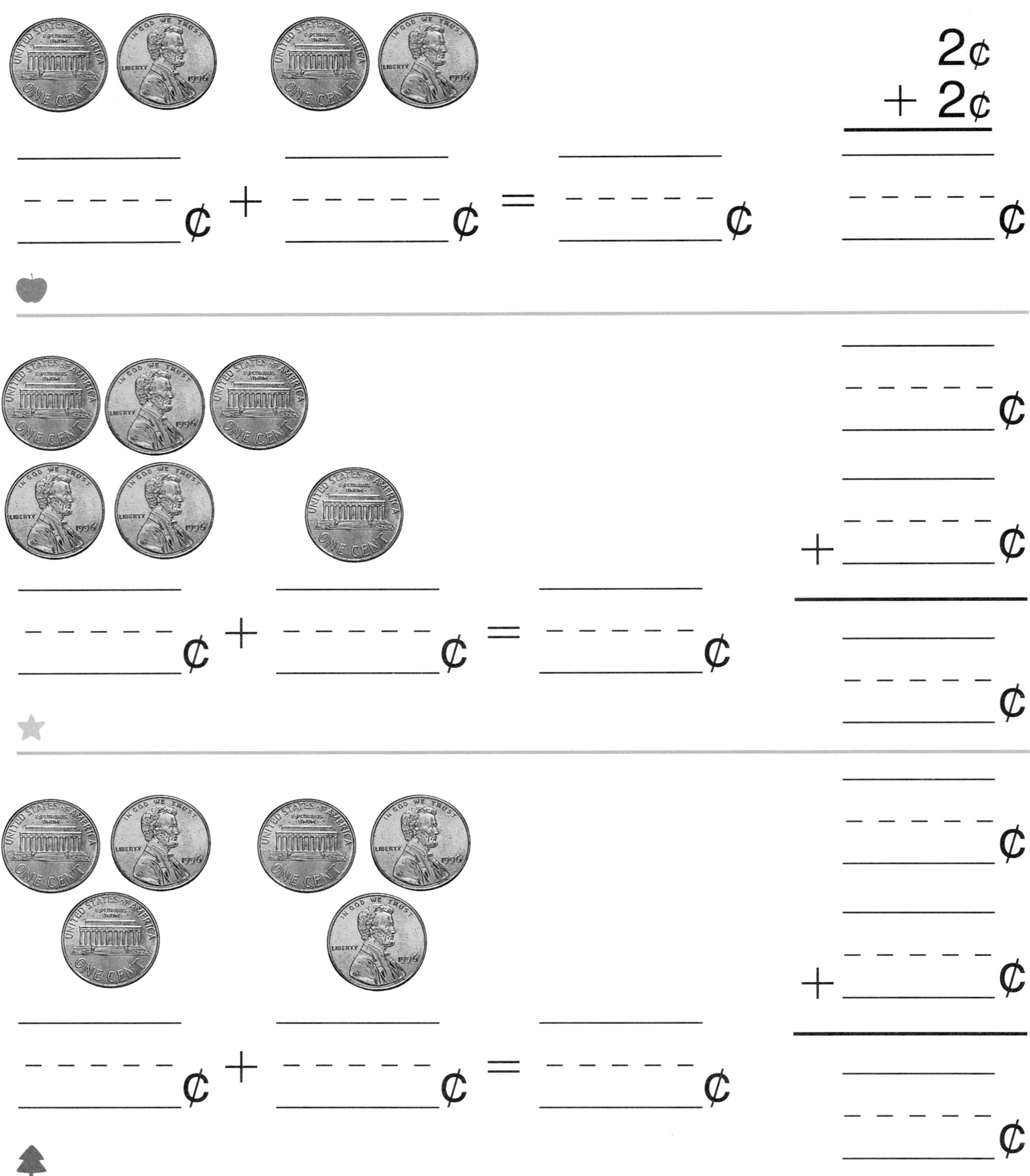

Add the money amounts. Write the addition two ways.

Name ____________________

4 and 1 is ____

____ and ____ is ____

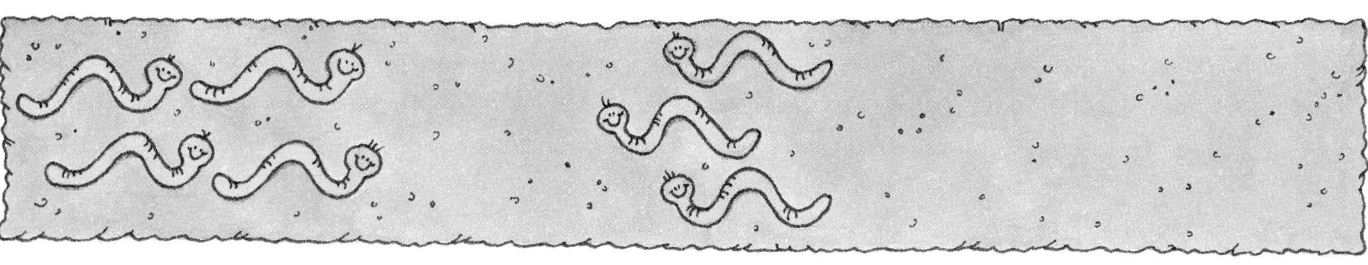

4 + 3 = ____

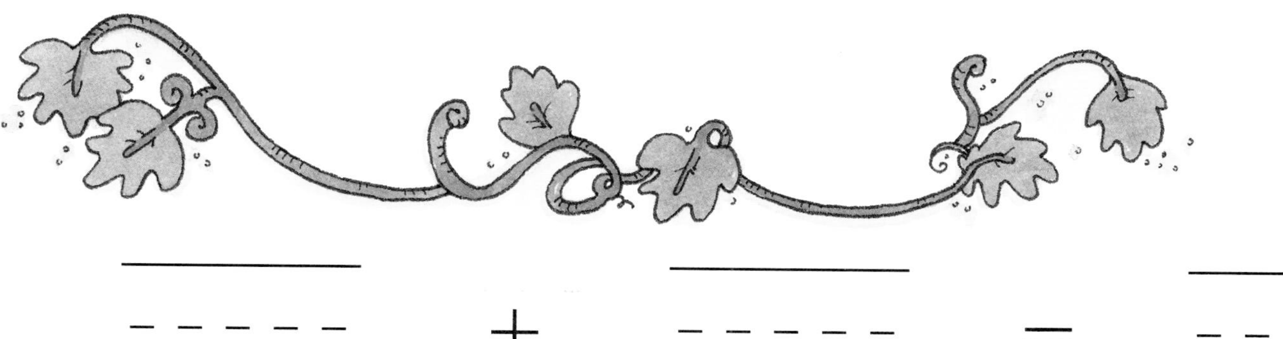

____ + ____ = ____

- Tell how many in each group. Tell how many in all. Write the number.
- Count on to find how many there are in all. Write the numbers.
- Add to find how many there are in all. Write the sum.
- Listen to the story. Draw pictures to solve the problem. Write the addition sentence.

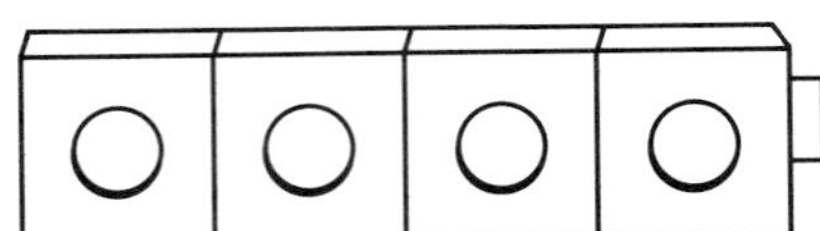

____ + ____ = ____

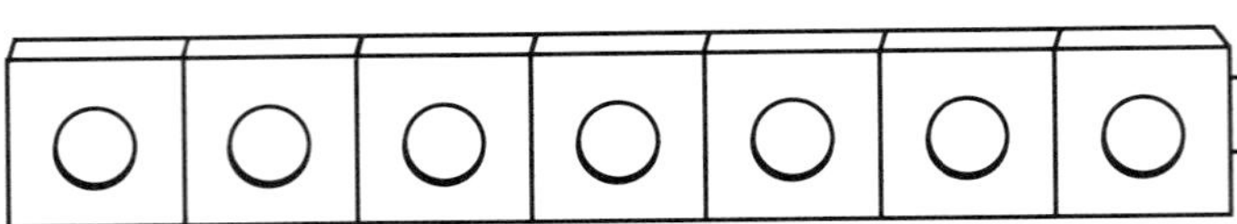

____ + ____ = ____

$$\begin{array}{r} 2 \\ +\ 2 \\ \hline \end{array}$$

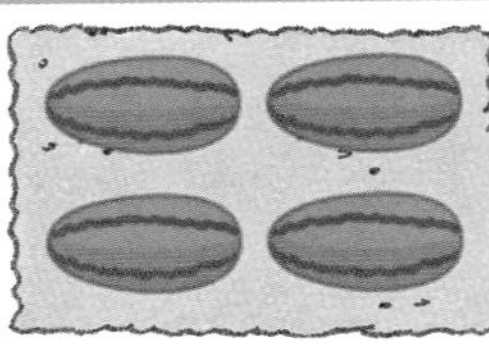

$$\begin{array}{r} 3 \\ +\ 1 \\ \hline \end{array}$$

$5 + 2 =$ ____

$$\begin{array}{r} 5 \\ +\ 2 \\ \hline \end{array}$$

2¢ + 1¢ = ____ ¢

$$\begin{array}{r} 2¢ \\ +\ 1¢ \\ \hline \end{array}$$

____ ¢

Use 2 different colors of cubes to make the number. Color. Write the addition sentence.
Look at each picture. Add. Add. You can use cubes.
Add the money amounts. Write the sum.

CHAPTER

13 Subtraction Concepts

theme
Flying Friends

Use the Data

How many dragonflies are flying away?

What You Will Learn

In this chapter you will learn how to:

- Show the meaning of subtraction.
- Explore subtraction in horizontal and vertical forms.
- Find differences from 10.
- Choose the operation to solve problems.

Dear Family,
In Chapter 13 I will learn about subtraction. Here are new vocabulary words and an activity that we can do together.

Kitchen Subtraction

- Put up to 10 spoons on a paper plate and have your child count them.
- Take some of the spoons off the plate.

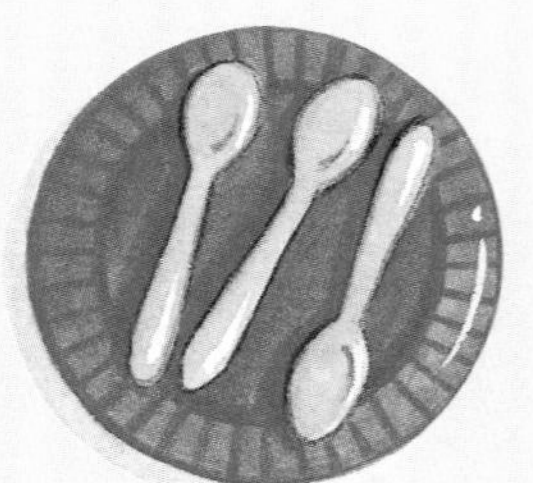

- Have your child count how many spoons are left on the plate.

use

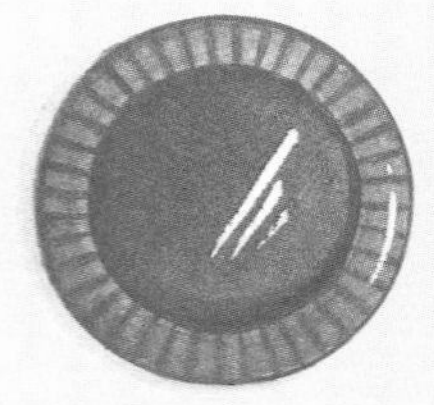

paper plate

and pasta

or spoons

Math Words

subtract

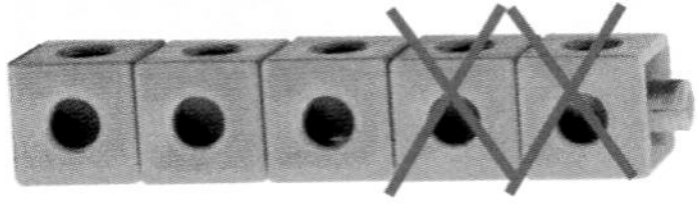

5 - 2 = 3

$$\begin{array}{r} 5 \\ -2 \\ \hline 3 \end{array}$$

minus —
5 **minus** 2 equals 3

equals =
5 minus 2 **equals** 3

difference
5 - 2 = 3
↑
The difference tells how many are left.

Additional activities at www.mhschool.com/math

5 take away 2 is 3

4 take away 1 is ____

____ take away ____ is ____

Tell how many there are in all. Tell how many fly away. Tell how many are left. Write the numbers.

Math at Home: Your child is learning how to separate groups.
Activity: Display up to 5 small objects. Then take away some. Ask your child to tell how many are left.

4 take away 2 is ______

5 take away 1 is ______

3 take away 1 is ______

4 take away 3 is ______

______ take away ______ is ______

Tell how many there are in all. Tell how many fly away. Tell how many are left. Write the numbers.

Name ____________________

6 take away 1 is 5

7 take away 1 is ____

____ take away ____ is ____

Tell how many there are in all. Tell how many are flying away. Tell how many are left. Write the numbers.

Math at Home: Your child subtracted by taking away 1.
Activity: Show your child up to 10 small objects and then take 1 away. Ask your child to tell how many are left.

4 take away 1 is 

9 take away 1 is 

3 take away 1 is 

5 take away 1 is 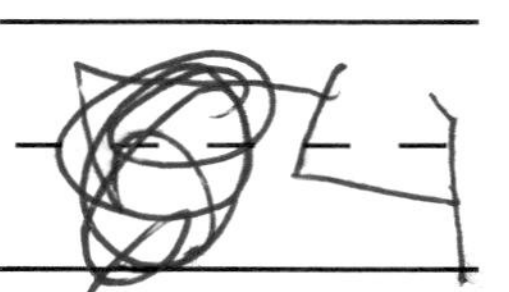

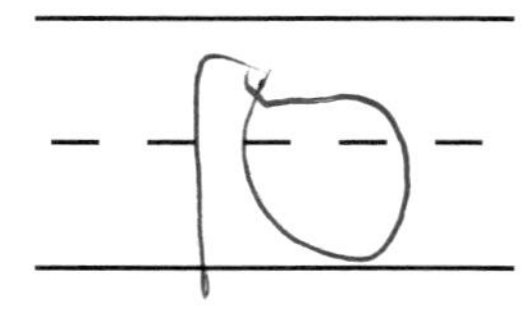 take away  is 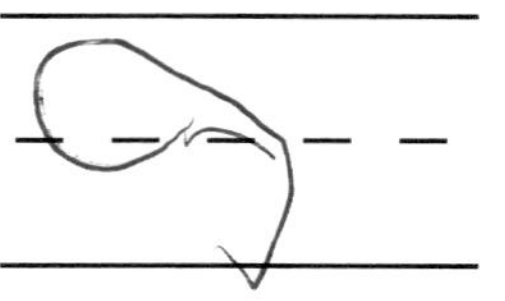

Tell how many there are in all. Tell how many are taken away. Tell how many are left. Write the numbers.

Name ________________________________

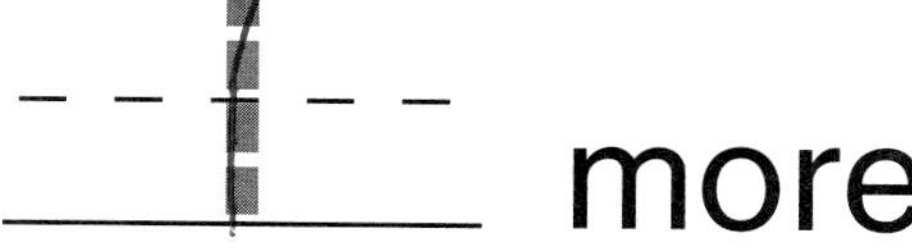

more

more

more

Look at the pictures. Draw lines to compare. Write how many more.

Math at Home: Your child subtracted by comparing numbers.
Activity: Show your child a group of up to 5 small objects. Then show another group with fewer objects. Ask your child to tell many more are in the first group.

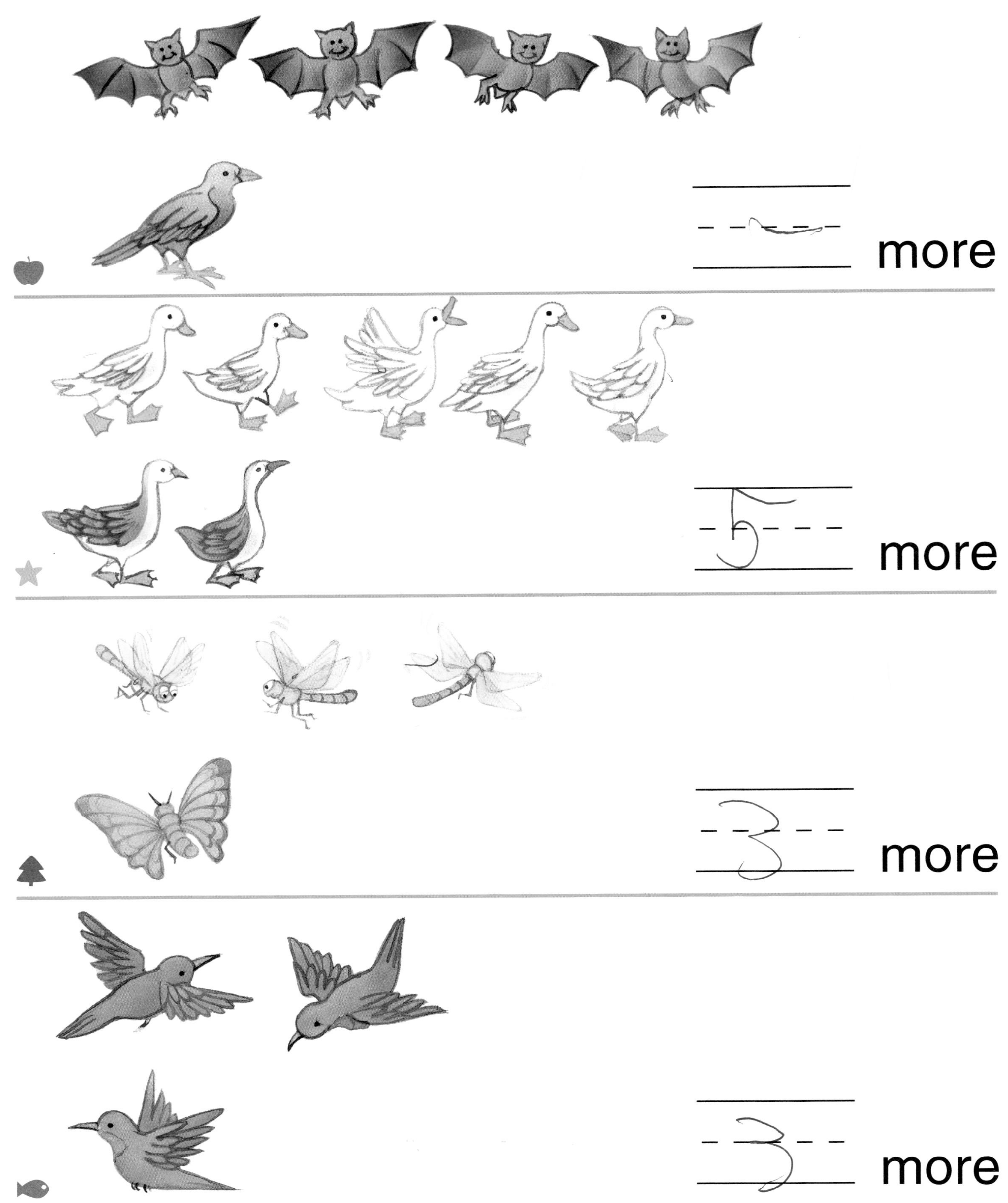

Look at the pictures. Draw lines to compare. Write how many more.

Name ______________________

Math Words
subtract
minus
equal
difference

Say these words.

5 − 1 = 4

4 − 2 = ____

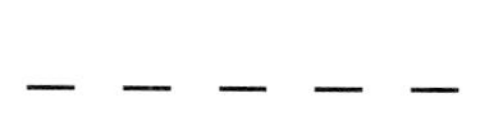

3 − 1 = ____

Tell a story about each picture. Subtract to find how many are left. Write each difference.

Math at Home: Your child learned subtraction and the - and = signs.
Activity: Ask your child to tell you what each sign means.

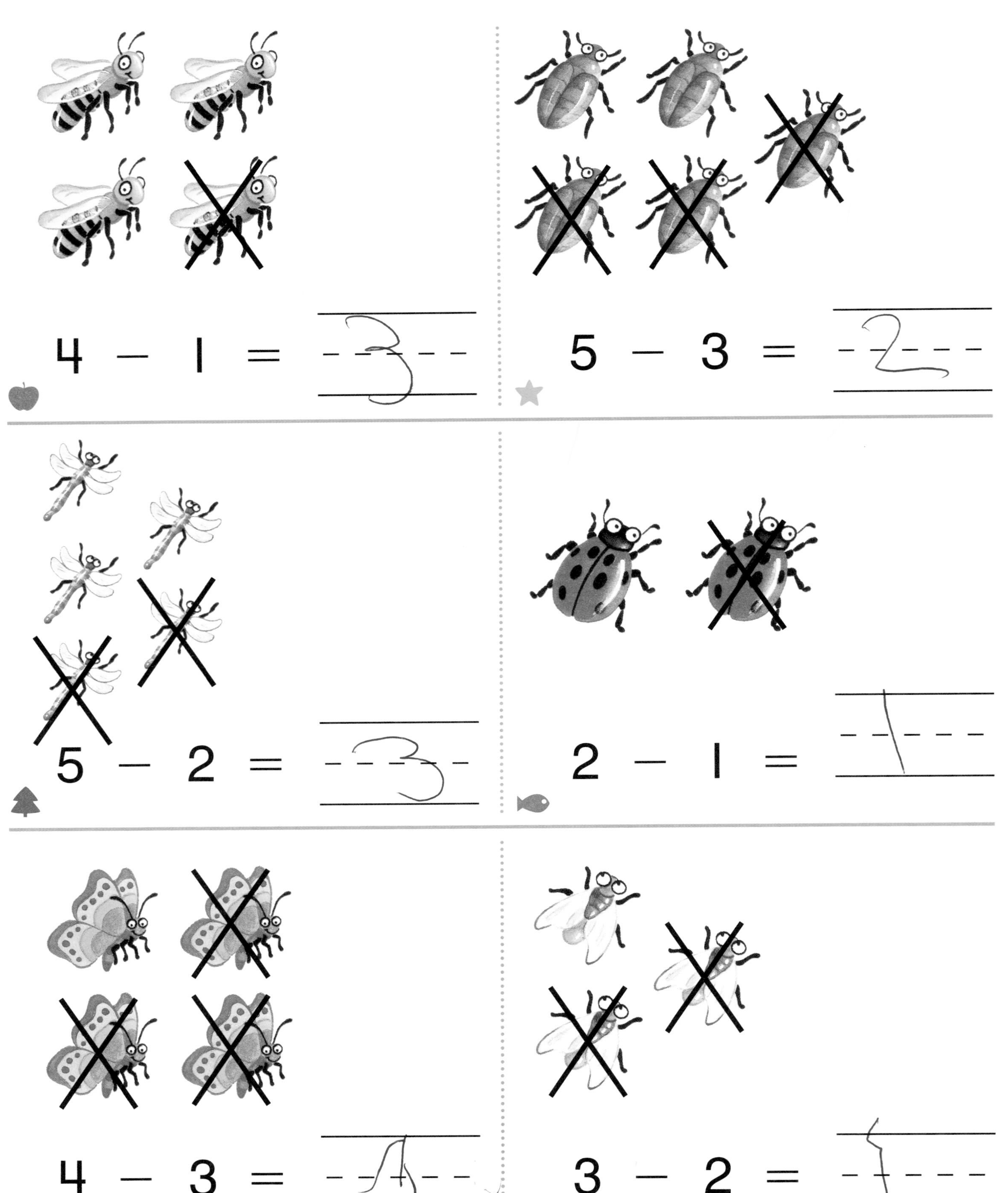

Tell a story about each picture. Subtract to find how many are left. Write each difference.

Name ______________________________

Choose the Operation

2 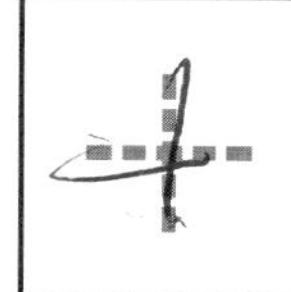 1 = 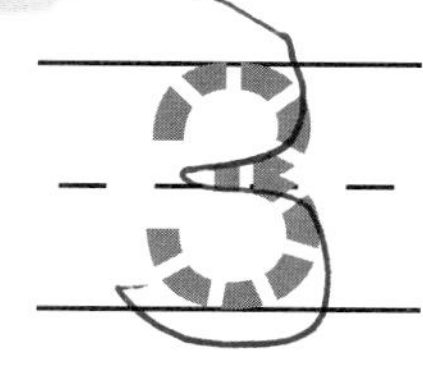

5 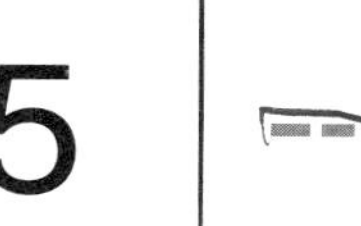 1 = 4

Listen to the story. Do you need to add or subtract? Write the sign. Write the answer.

Math at Home: Your child chose whether to add or subtract to solve the problem.
Activity: Make up a story problem such as: There are 4 red wagons. 1 is taken away. How many wagons are left? Ask your child whether they will need to add or subtract.

3 ☐ 2 = ____

5 ☐ 2 = ____

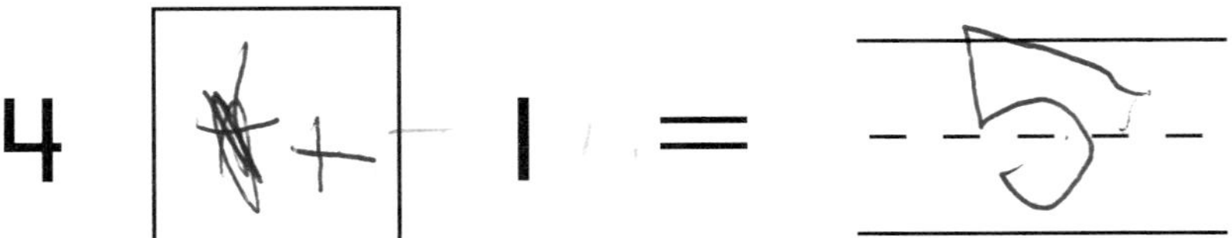

4 ☐ 1 = ____

Listen to the story. Do you need to add or subtract? Write the sign. Write the answer.

Name ____________________

4 − 1 =

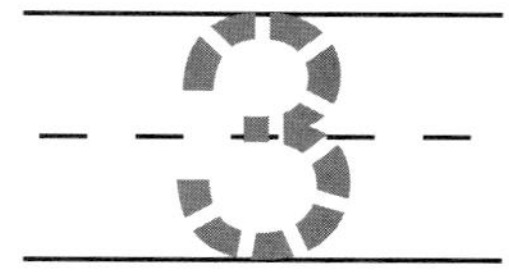

4 − 2 =

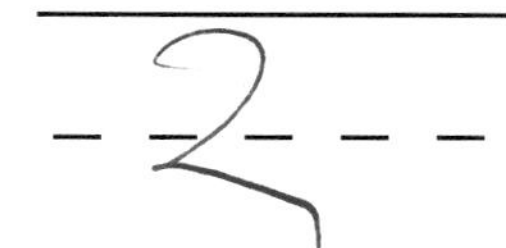

5 − 2 =

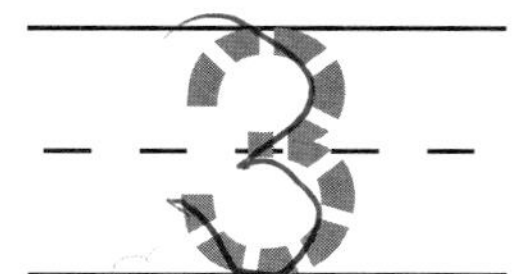

5 − 3 =

Look at the cubes to subtract. Write the difference.

Math at Home: Your child used cubes to find differences from 4 and 5.
Activity: Show your child 4 or 5 small objects. Take some objects away. Ask your child to tell how many are left.

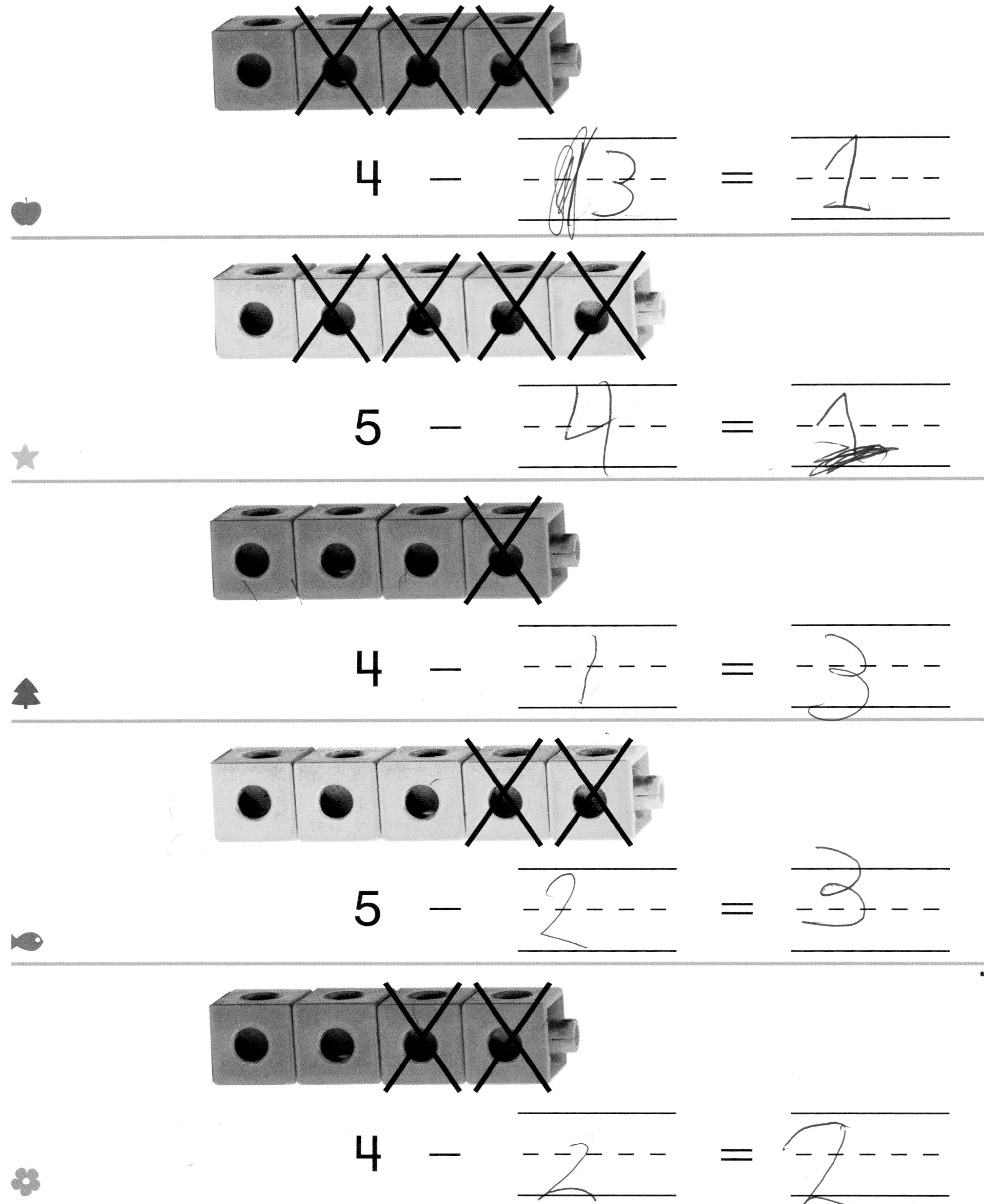

Look at the cubes. Write the subtraction sentence.

Name________________________________

Subtracting from 6 and 7

6 − 1 = 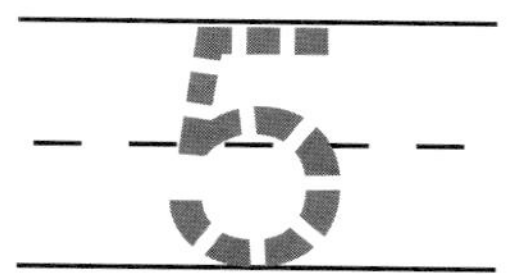

6 − 2 = 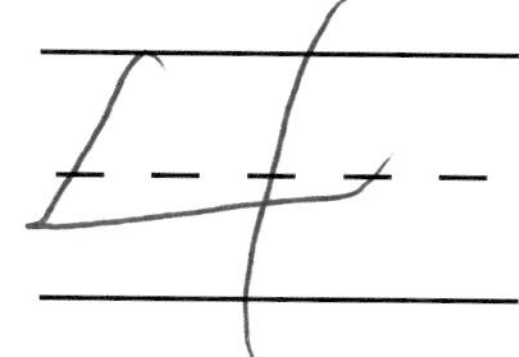

7 − 2 = 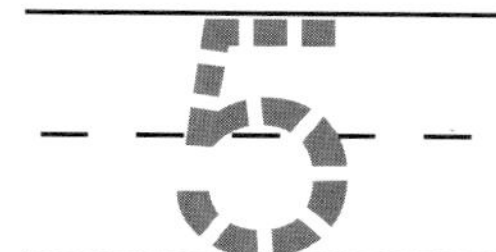

7 − 3 = 

Look at the cubes to subtract. Write the difference.

Math at Home: Your child used cubes to find differences from 6 and 7.
Activity: Show your child 6 or 7 small objects. Take some objects away. Ask your child to tell how many are left.

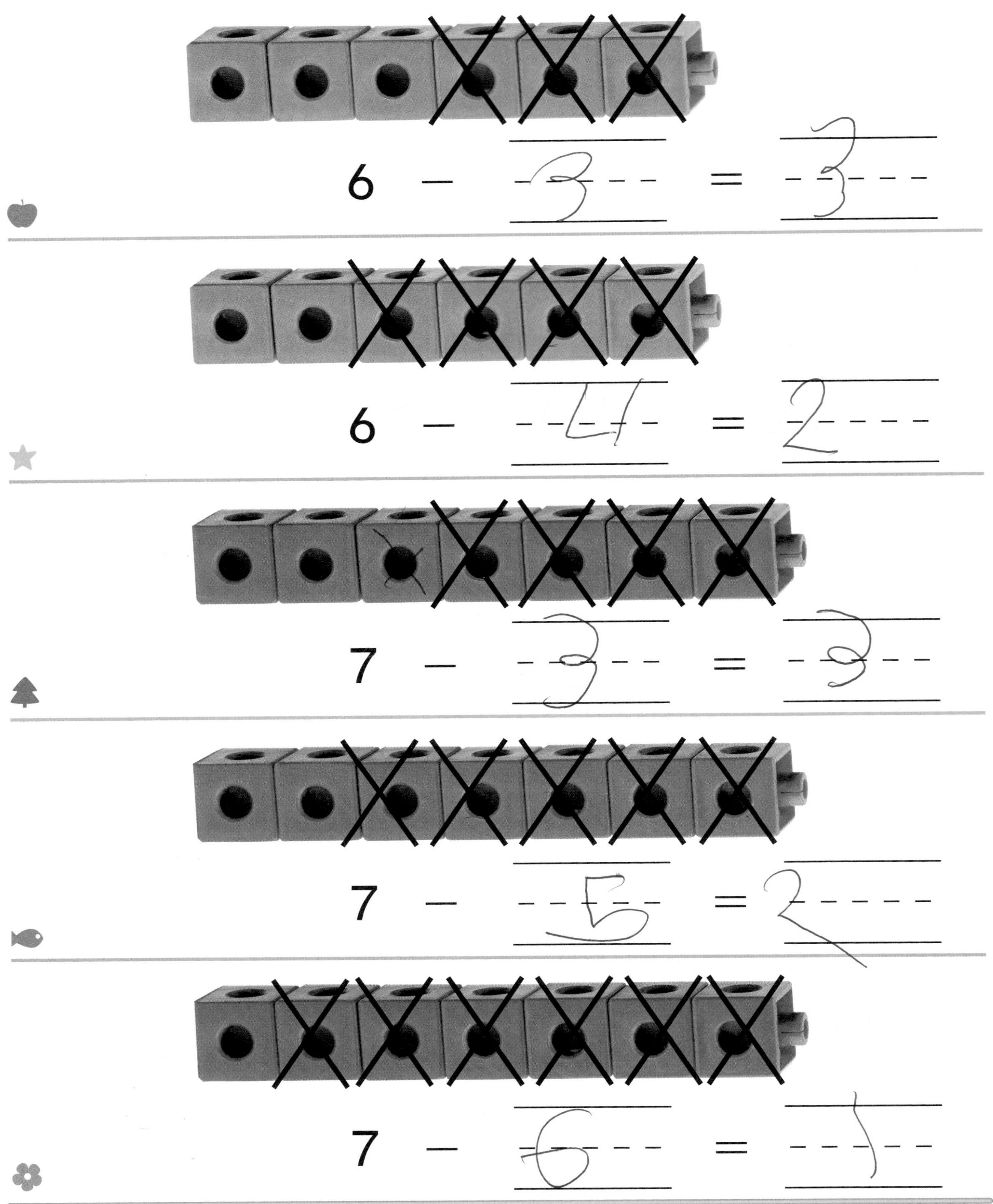

Look at the cubes. Write the subtraction sentence.

Name ______________________________

8 − 1 = 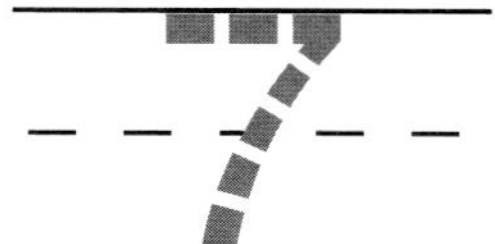

8 − 2 = 

9 − 4 = 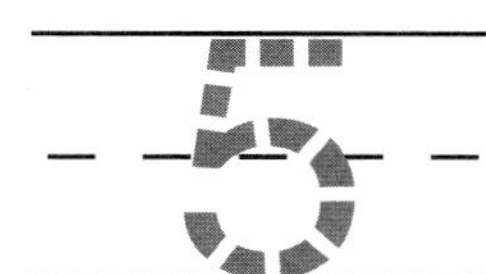

9 − 2 = 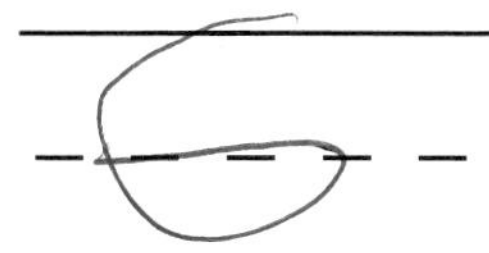

Look at the cubes to subtract. Write the difference.

Math at Home: Your child used cubes to find differences from 8 and 9.
Activity: Show your child 8 or 9 small objects. Take some objects away. Ask your child to tell how many are left.

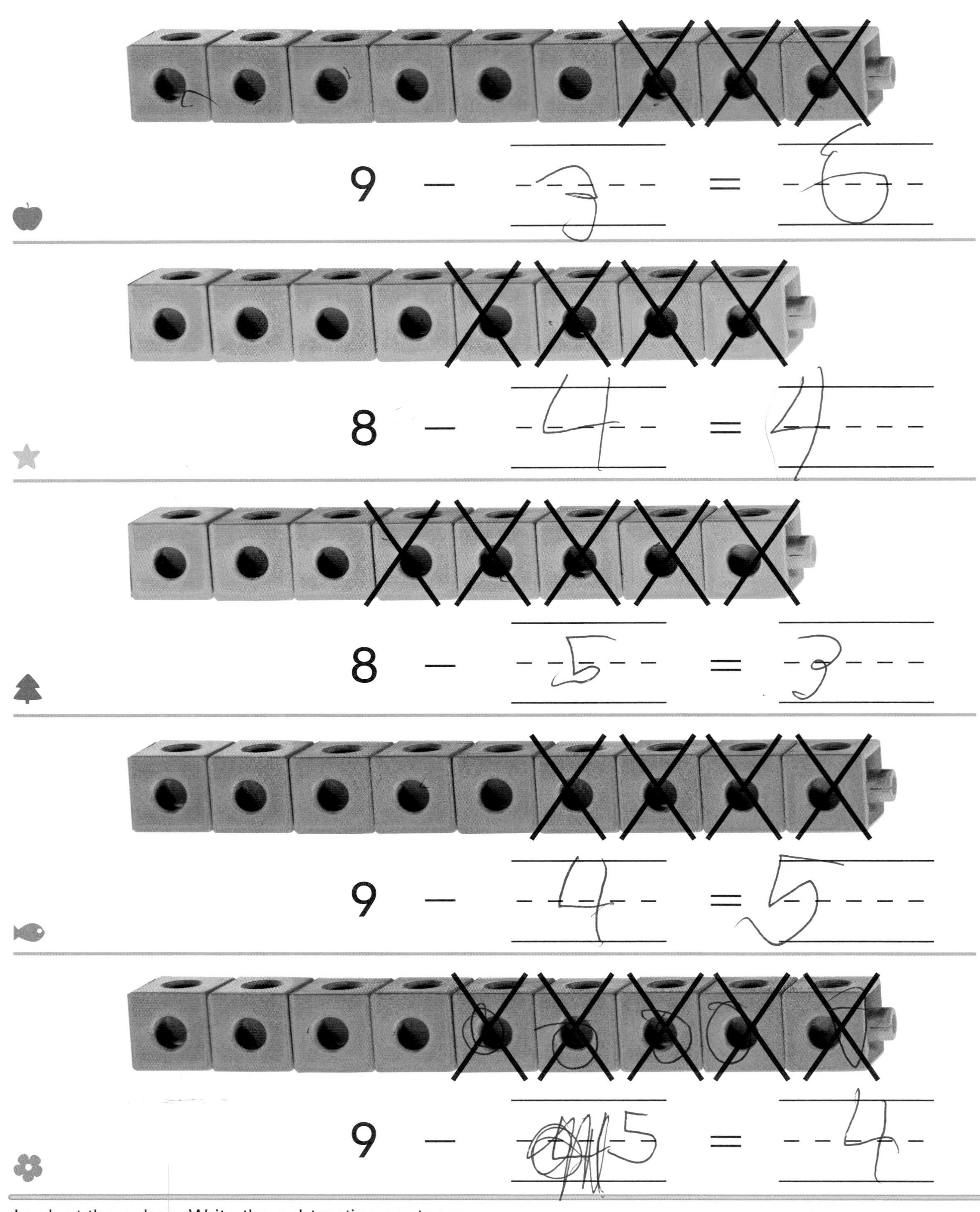

Look at the cubes. Write the subtraction sentence.

Name ______________________________

10 − 1 = 9

10 − 2 = 8

10 − 3 = 7

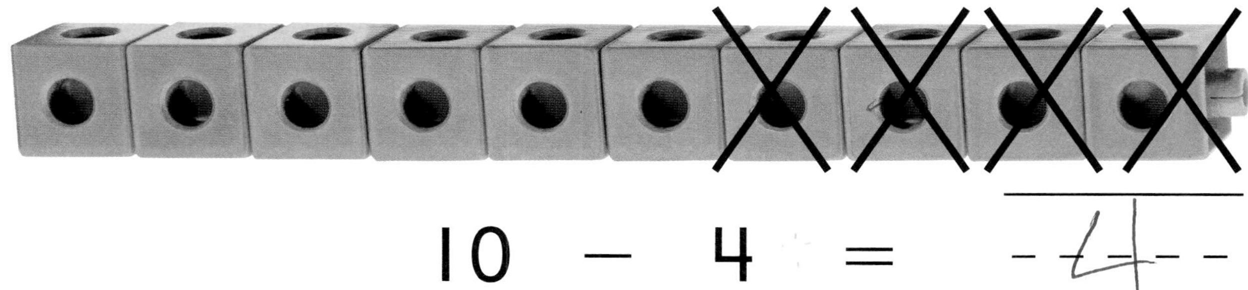

10 − 4 = 4

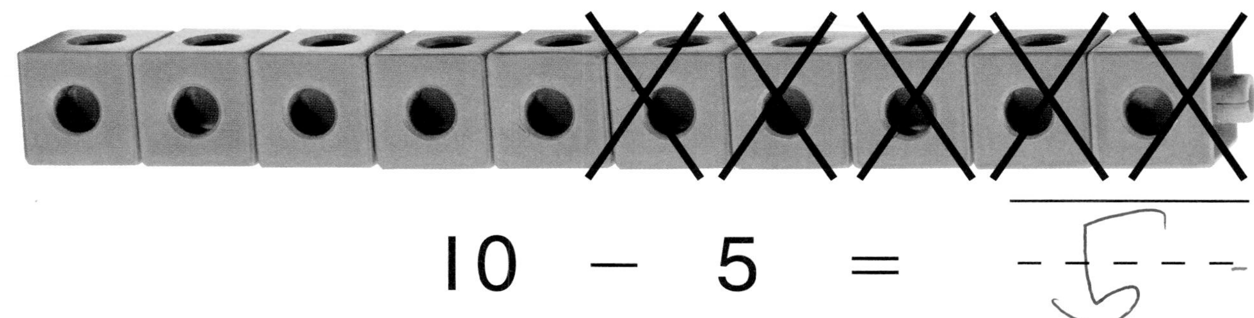

10 − 5 = 5

Look at the cubes to subtract. Write the difference.

Math at Home: Your child used cubes to find differences from 10.
Activity: Show your child 10 small objects. Take some objects away. Ask your child to tell how many are left.

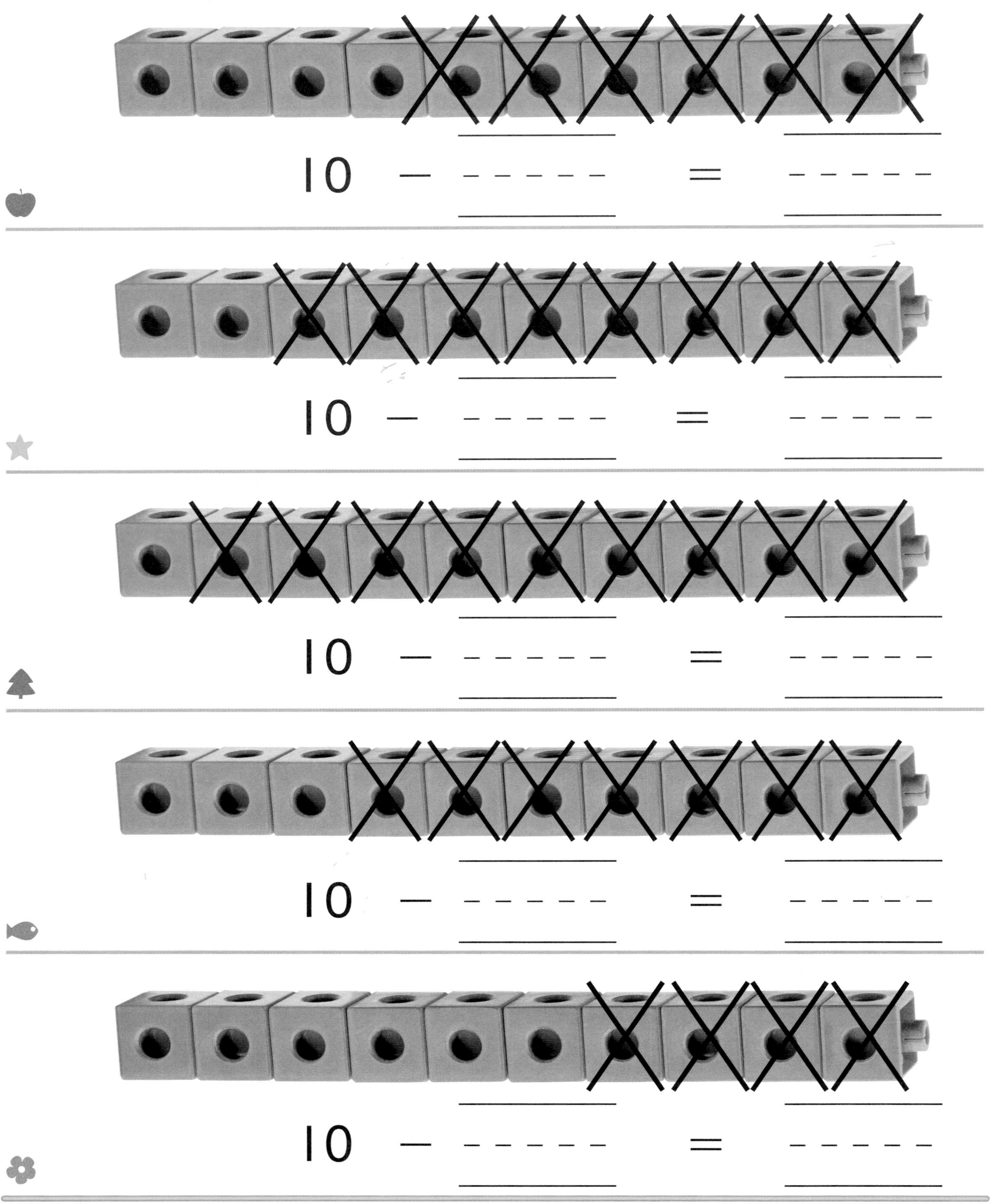

Look at the cubes. Write the subtraction sentence.

Name ____________________

$$\begin{array}{r} 5 \\ -\ 1 \\ \hline 4 \end{array}$$

$$\begin{array}{r} 5 \\ -\ 2 \\ \hline 3 \end{array}$$

$$\begin{array}{r} 4 \\ -\ 1 \\ \hline \end{array}$$

$$\begin{array}{r} 3 \\ -\ 2 \\ \hline \end{array}$$

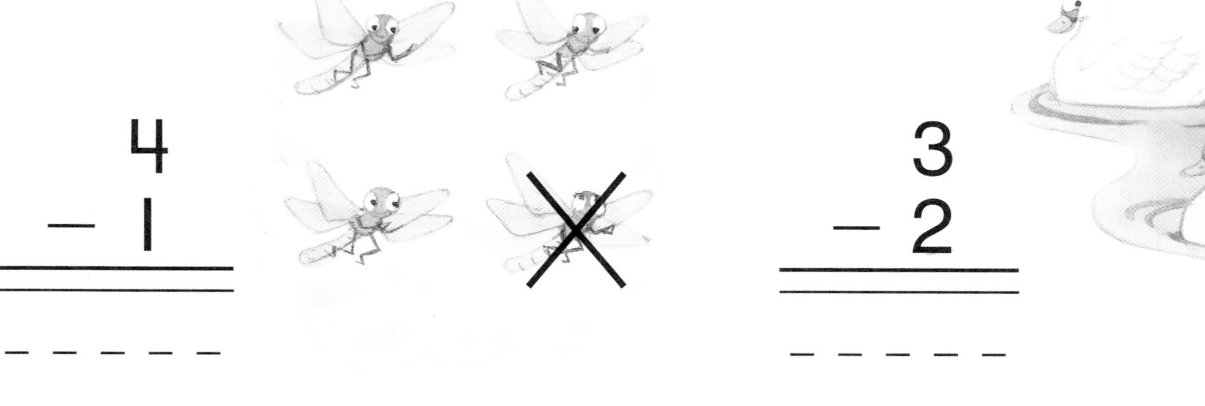

Look at each picture. Subtract.

Math at Home: Your child is learning vertical subtraction.
Activity: Write a vertical subtraction problem from a number up to ten and ask your child to tell you the answer.

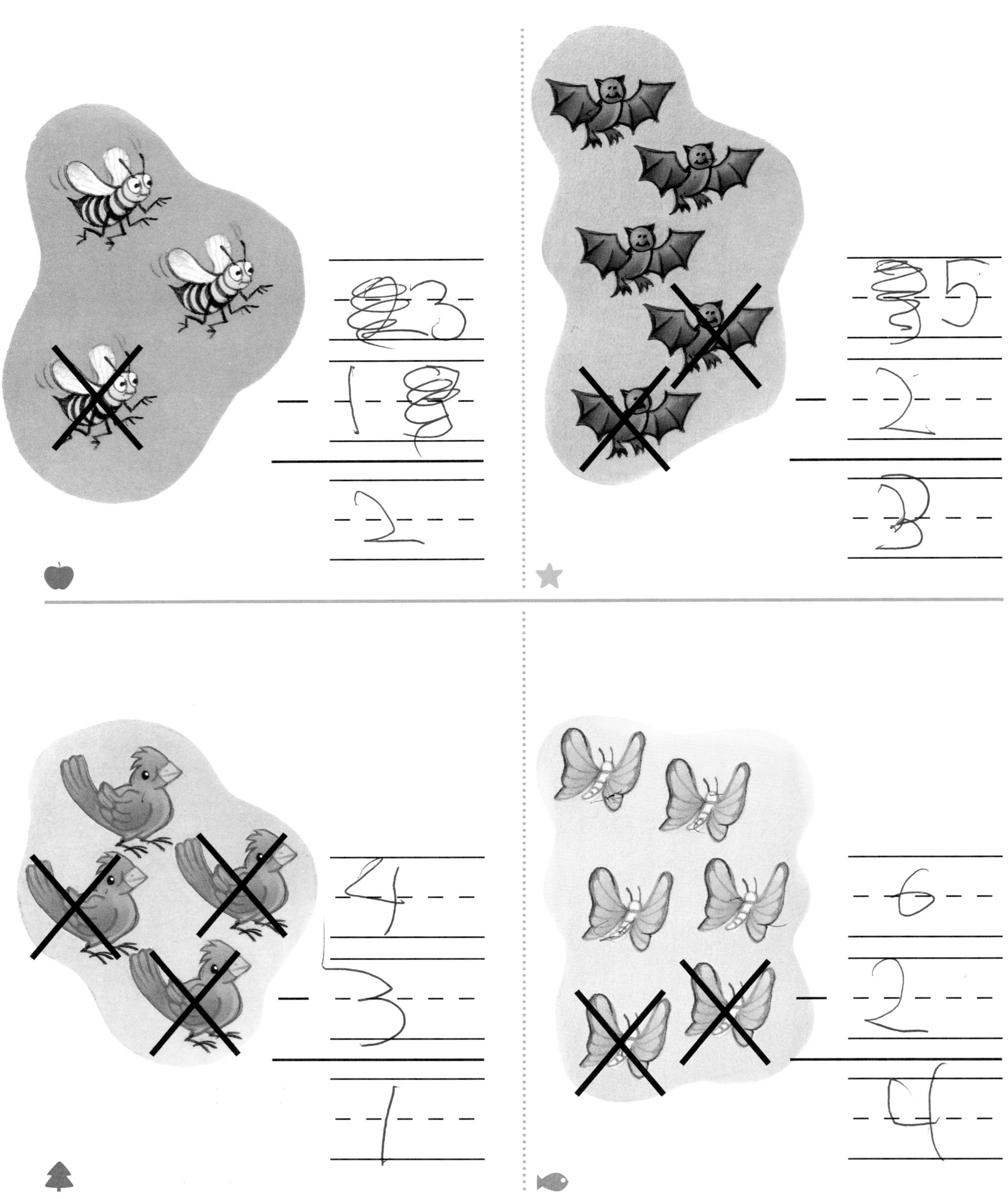

Look at each picture. Write the subtraction.

Name ______________________

5 − 2 = 3

$$\begin{array}{r} 5 \\ -\ 2 \\ \hline 3 \end{array}$$

4 − 1 = ____

$$\begin{array}{r} 4 \\ -\ 1 \\ \hline \end{array}$$

4 − 2 = ____

$$\begin{array}{r} 4 \\ -\ 2 \\ \hline \end{array}$$

6 − 2 = ____

$$\begin{array}{r} 6 \\ -\ 2 \\ \hline \end{array}$$

3 − 1 = ____

$$\begin{array}{r} 3 \\ -\ 1 \\ \hline \end{array}$$

Subtract. You can use cubes.

Math at Home: Your child practiced subtracting numbers horizontally and vertically.
Activity: Point to a problem on this page and ask your child to tell you how to find how many are left.

$\begin{array}{r}4\\-3\\\hline\end{array}$	$\begin{array}{r}2\\-1\\\hline\end{array}$	$\begin{array}{r}6\\-1\\\hline\end{array}$	$\begin{array}{r}5\\-1\\\hline\end{array}$

$\begin{array}{r}1\\-1\\\hline\end{array}$	$\begin{array}{r}3\\-2\\\hline\end{array}$	$\begin{array}{r}8\\-1\\\hline\end{array}$	$\begin{array}{r}6\\-3\\\hline\end{array}$

$\begin{array}{r}2\\-2\\\hline\end{array}$	$\begin{array}{r}5\\-4\\\hline\end{array}$	$\begin{array}{r}7\\-2\\\hline\end{array}$	$\begin{array}{r}3\\-1\\\hline\end{array}$

$\begin{array}{r}7\\-1\\\hline\end{array}$	$\begin{array}{r}6\\-4\\\hline\end{array}$	$\begin{array}{r}3\\-3\\\hline\end{array}$	$\begin{array}{r}5\\-3\\\hline\end{array}$

Subtract. You can use cubes.

Name ____________________

6¢ − 2¢ = 4¢

$$\begin{array}{r} 6¢ \\ -\ 2¢ \\ \hline \_\_\_¢ \end{array}$$

5¢ − 2¢ = ___¢

$$\begin{array}{r} 5¢ \\ -\ 2¢ \\ \hline \_\_\_¢ \end{array}$$

4¢ − 1¢ = ___¢

$$\begin{array}{r} 4¢ \\ -\ 1¢ \\ \hline \_\_\_¢ \end{array}$$

5¢ − 3¢ = ___¢

$$\begin{array}{r} 5¢ \\ -\ 3¢ \\ \hline \_\_\_¢ \end{array}$$

Subtract the money amounts. Write the difference.

Math at Home: Your child subtracted pennies.
Activity: Show your child a group of pennies that total 10¢ or less. Ask your child to tell you how many pennies in all. Then take some pennies away and have your child tell how many are left.

$$\begin{array}{r} 4¢ \\ -\ 2¢ \\ \hline \end{array}$$

____¢ − ____¢ = ____¢ ____¢

$$\begin{array}{r} 6¢ \\ -\ 1¢ \\ \hline \end{array}$$

____¢ − ____¢ = ____¢ ____¢

$$\begin{array}{r} 6¢ \\ -\ 3¢ \\ \hline \end{array}$$

____¢ − ____¢ = ____¢ ____¢

$$\begin{array}{r} 5¢ \\ -\ 1¢ \\ \hline \end{array}$$

____¢ − ____¢ = ____¢ ____¢

Subtract the money amounts. Write the difference.

Name ______________________

4 take away 2 is 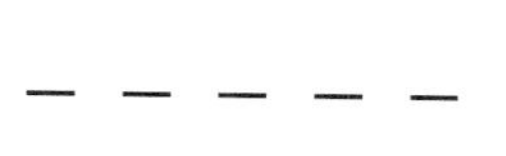

 more

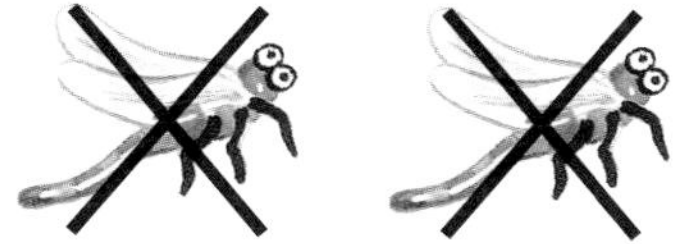

4 − 3 = ______

5 − 2 = ______

3  1 = 

- Tell how many in all. Tell how many fly away. Tell how many are left. Write the number.
- Look at the pictures. Draw lines to compare. Write how many more.
- Subtract to find how many are left. Write each difference.
- Listen to the story. Do you need to add or subtract? Write the sign. Write the answer.

5 − ______ = ______

8 − ______ = ______

$$\begin{array}{r} 6 \\ -\ 1 \\ \hline \end{array}$$

______

$$\begin{array}{r} 3 \\ -\ 2 \\ \hline \end{array}$$

______

4 − 2 = ______

$$\begin{array}{r} 4 \\ -\ 2 \\ \hline \end{array}$$

______

6¢ − 2¢ = 4¢

$$\begin{array}{r} 6¢ \\ -\ 2¢ \\ \hline \end{array}$$

______¢

Look at the cubes to subtract. Write the number sentence. Look at each picture. Subtract. Subtract. You can use cubes. Subtract the money amounts. Write the difference.

CHAPTER

14 Geometry and Fractions

theme

Snack Time

Use the Data

Which food could you share equally with a friend?

What You Will Learn

In this chapter you will learn how to:

- Sort solid figures.
- Relate plane figures and solid figures.
- Explore equal parts, equal groups, and halves.
- Act out situations to solve problems.

Dear Family,
In Chapter 14, I will learn about shapes and fractions. Here are new vocabulary words and an activity that we can do together.

Math Words

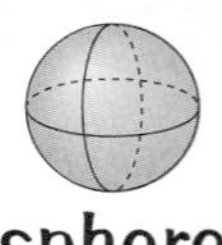
sphere

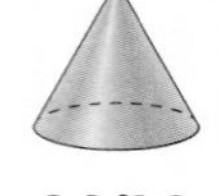
cone

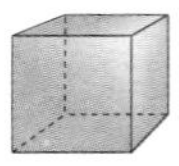
cube

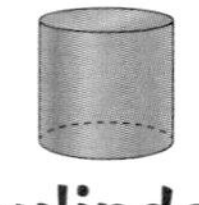
cylinder

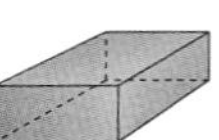
rectangular prism

circle

square

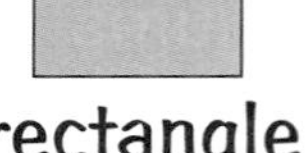
rectangle

triangle

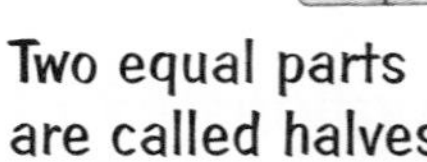
halves

Two equal parts are called halves.

Solid Sort

- Give your child a variety of objects. Have your child sort them into different groups.

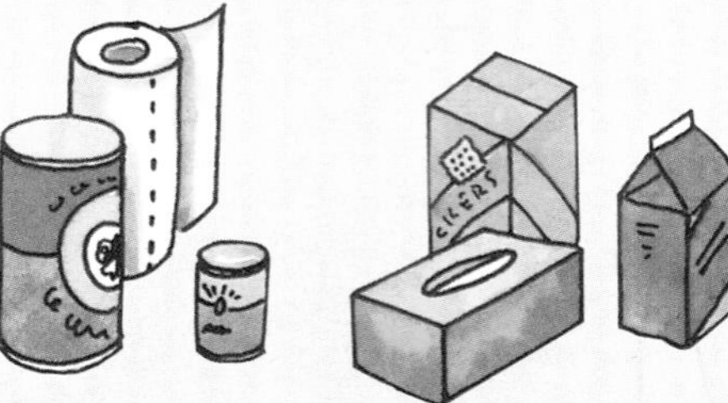

- Talk about how he or she sorted the objects.
- Ask your child to sort them a different way.

use

different shaped objects and containers

Additional activities at www.mhschool.com/math

Name ____________________

Say these words.

Math Words
rectangular prism
sphere
cube
cone
cylinder

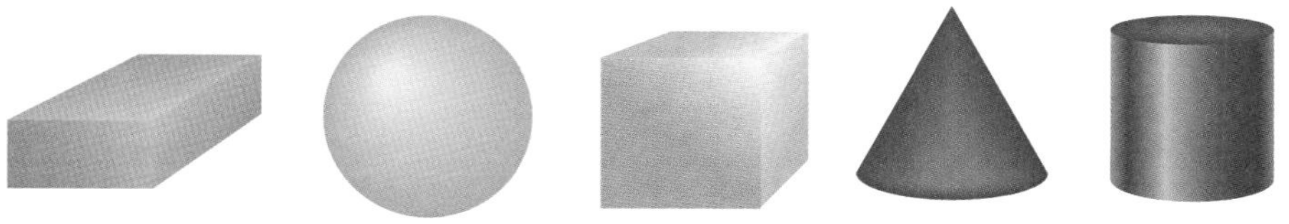

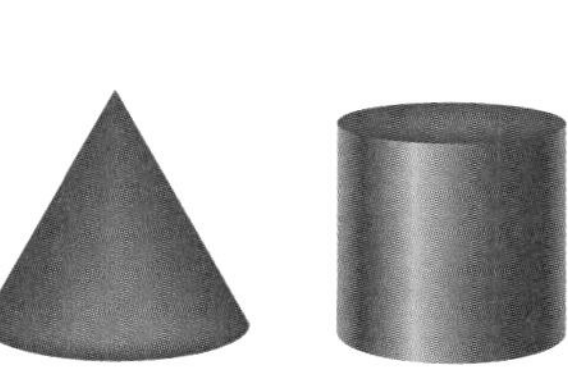

Draw a line from each solid figure to the snack that has the same shape.

Math at Home: Your child sorted objects according to shape.
Activity: Help your child find objects at home shaped like spheres, rectangular prisms, cones, cubes, and cylinders. Ask your child how the objects are alike.

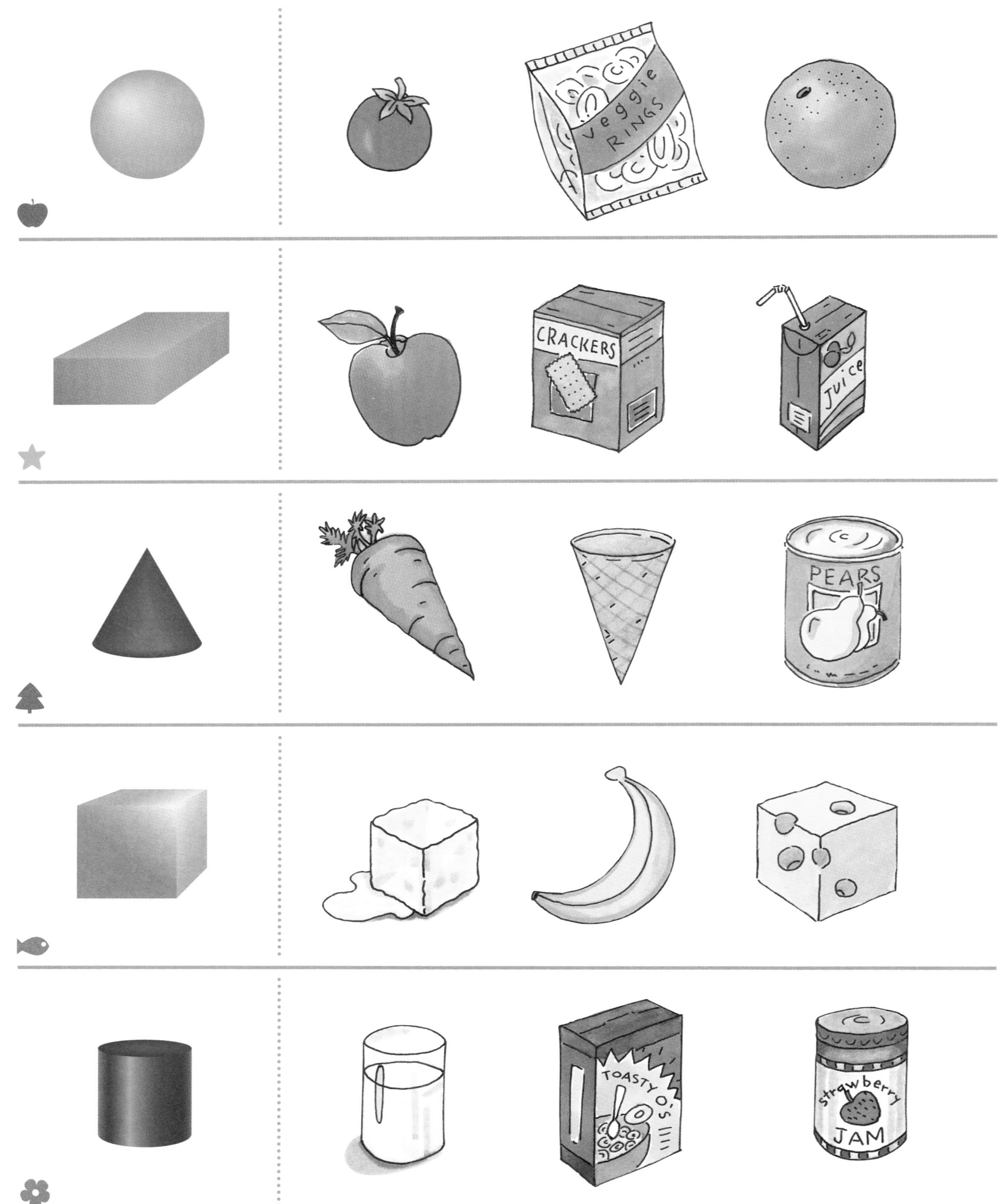

Circle the items in each row that have the same shape as the solid figure.

Name ____________________

Algebra & functions

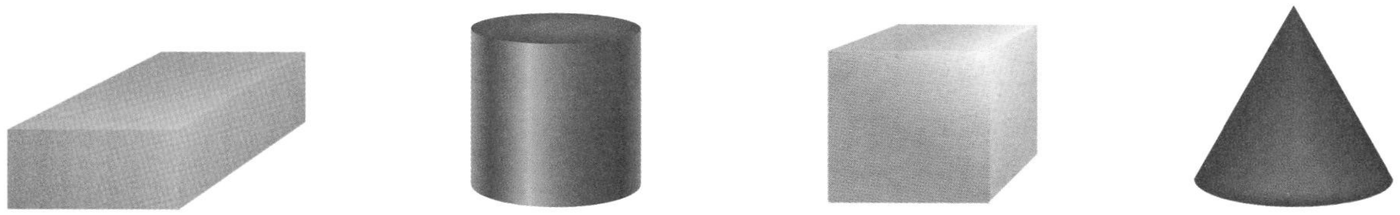

Trace solid figures to make shapes. Draw lines from each plane figure to the solid figure you traced to make it.

Math at Home: Your child made plane figures by tracing around the face of a solid figure.
Activity: Find solids at home like cereal boxes and soup cans. Have your child trace around the solids and tell you what plane figure he or she made.

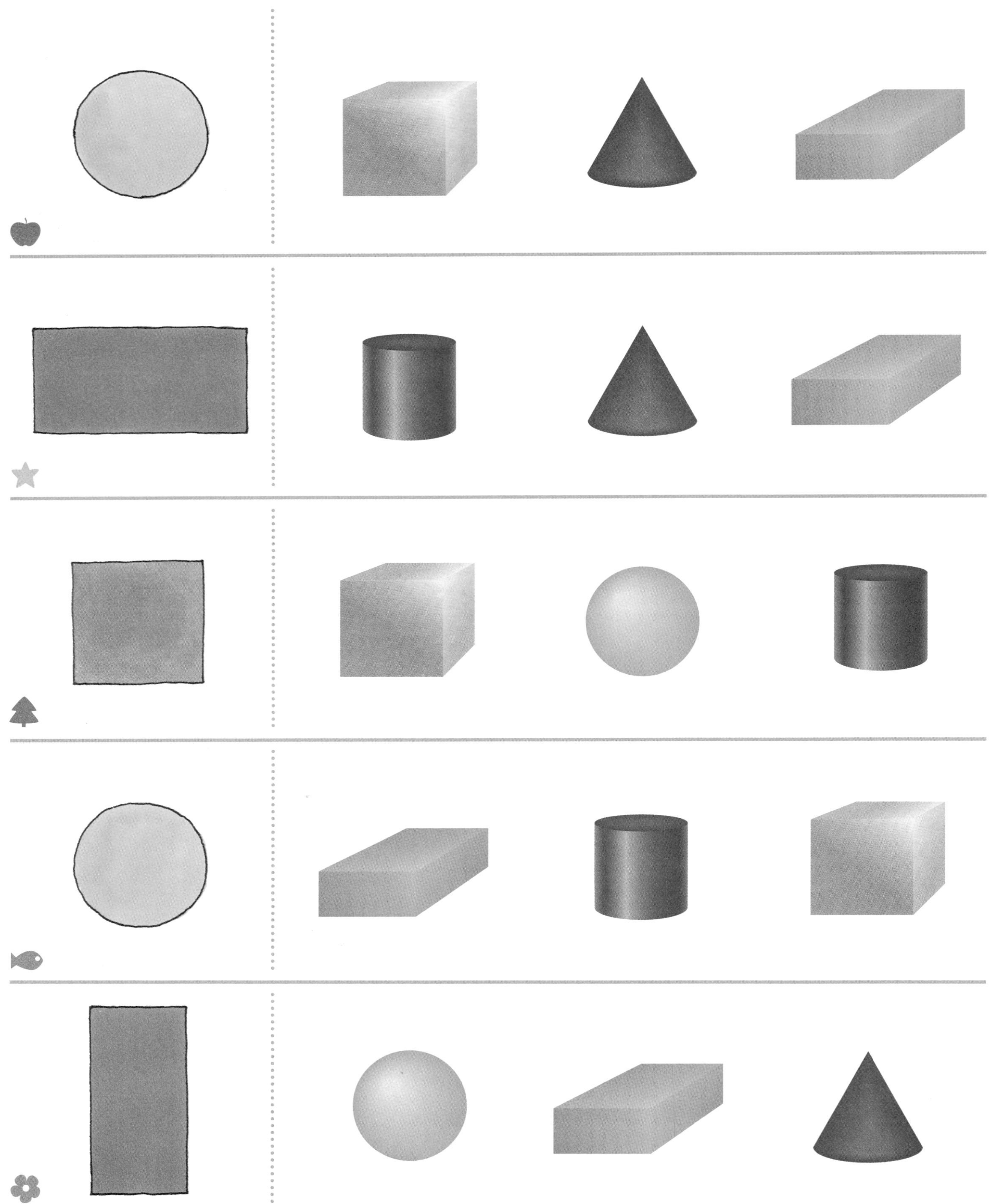

Circle the solid figure you can use to make each plane figure.

Name ______________________________

Algebra & functions

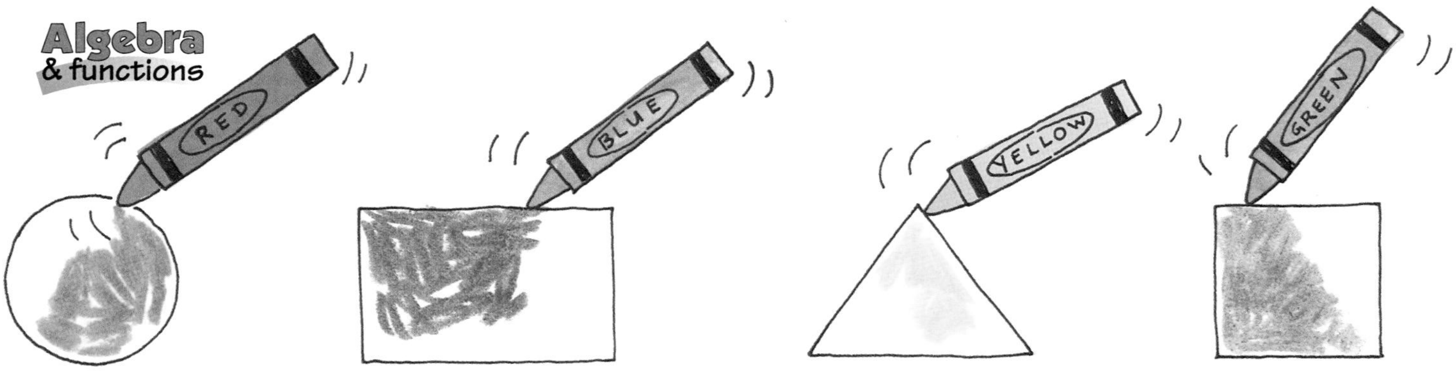

RED BLUE YELLOW GREEN

Color the circles red. Color the rectangles blue.
Color the triangles yellow. Color the squares green.

Math at Home: Your child identified circles, rectangles, triangles, and squares.
Activity: Ask your child to draw a picture of your family using circles, rectangles, triangles, and squares.

Circle the figures that have the same shape.

Name ____________________

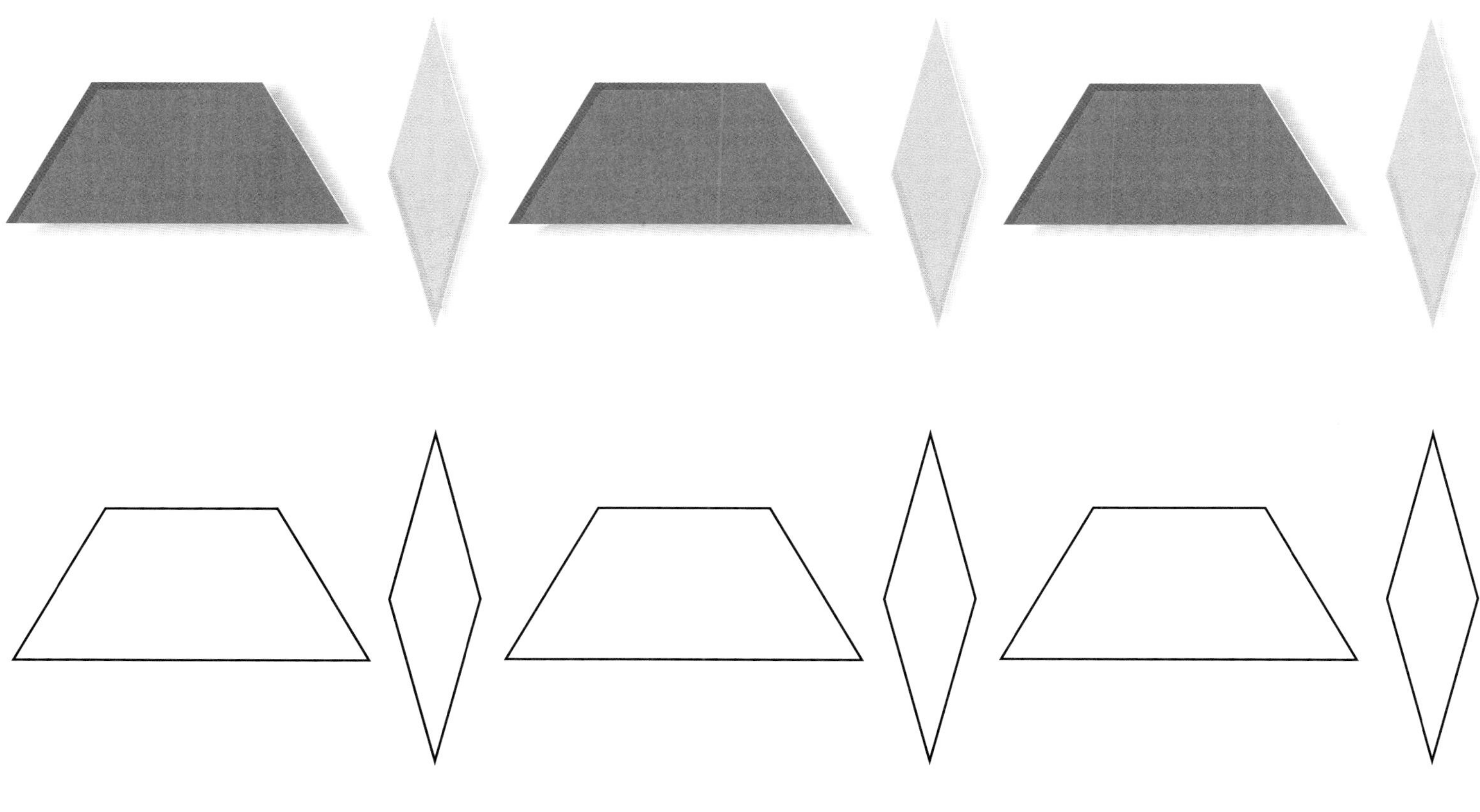

Use pattern blocks to copy the pattern. Color.

Math at Home: Your child copied and extended shape patterns.
Activity: Point to one of the patterns and ask your child how he or she knows what comes next.

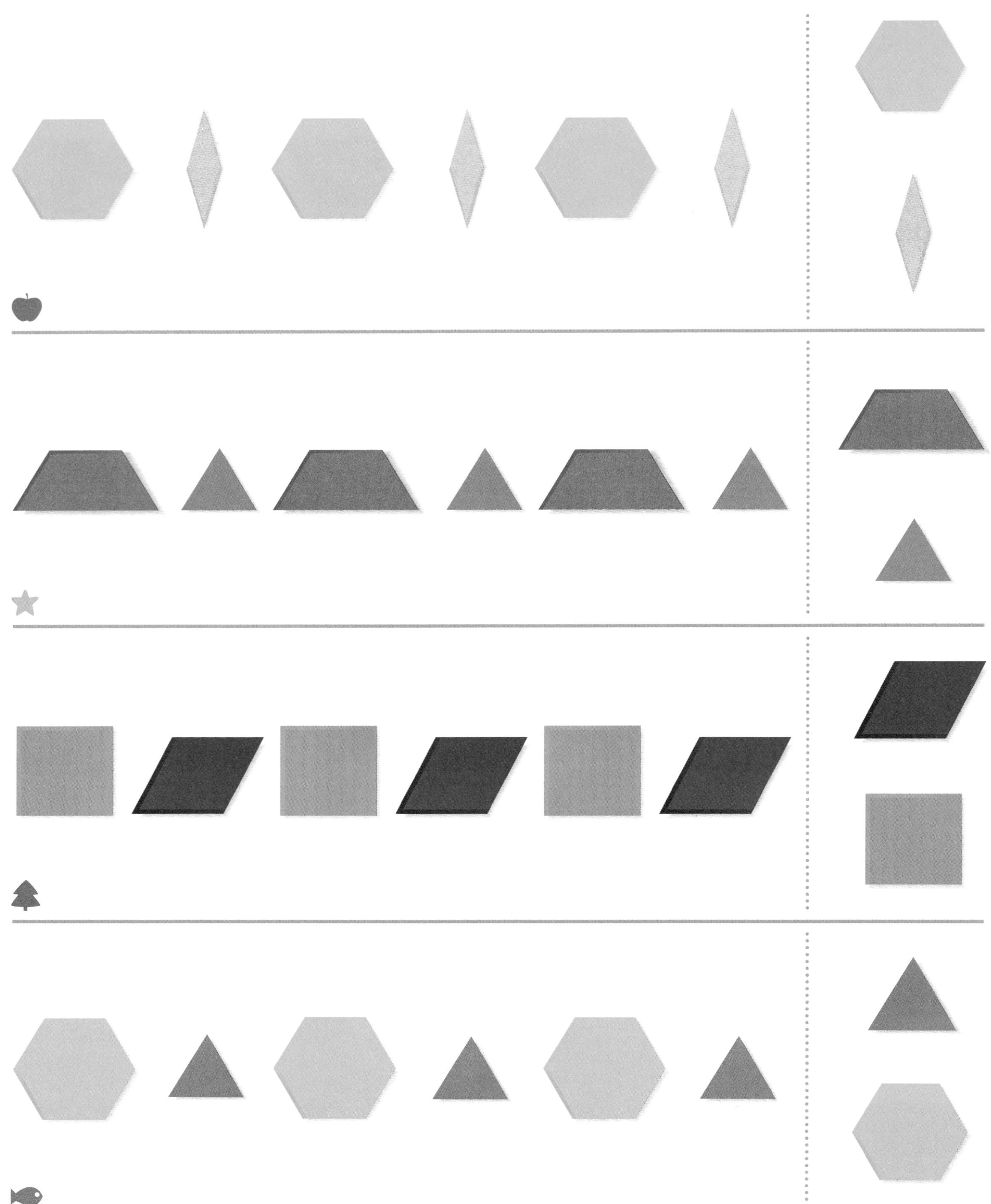

Circle what could come next in each pattern. Tell how you know.

Name______________________________

Act It Out

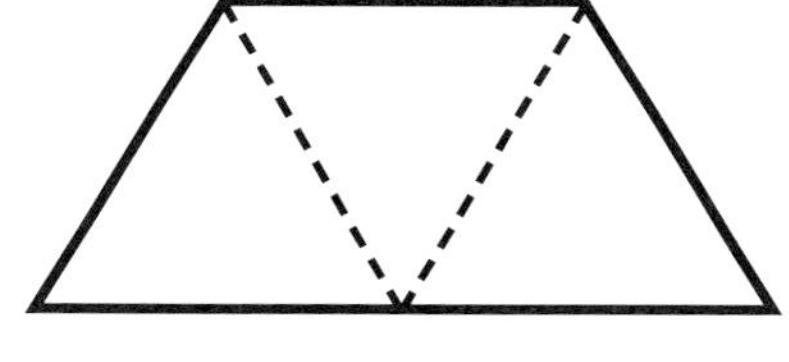

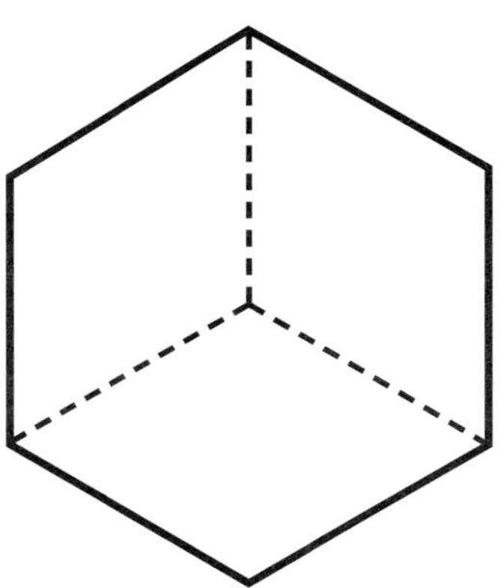

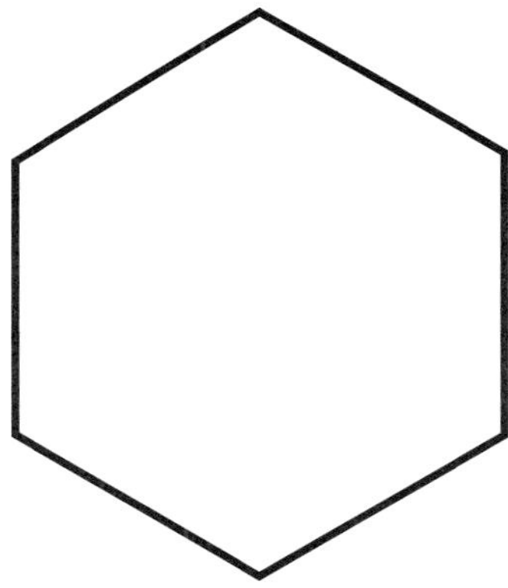

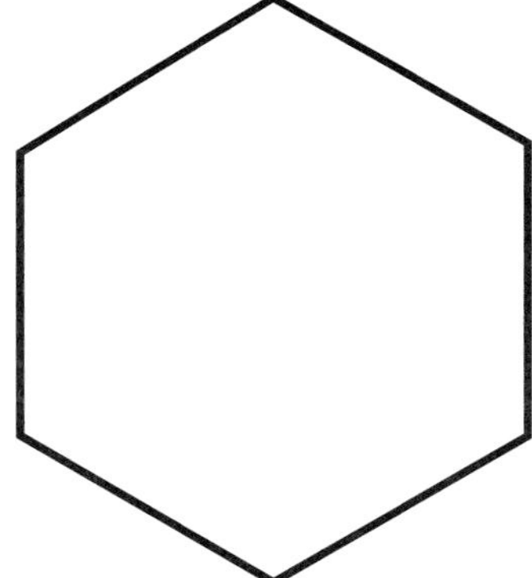

Use pattern blocks to make each shape in three different ways.
Color to show the pattern blocks you used.

Math at Home: Your child used pattern blocks to make shapes.
Activity: Cut out some paper squares, triangles, and rectangles. Ask your child to put them together to make different shapes.

Use pattern blocks to make each shape. Color to show the pattern blocks you used.

Name ______________________________

equal parts

unequal parts

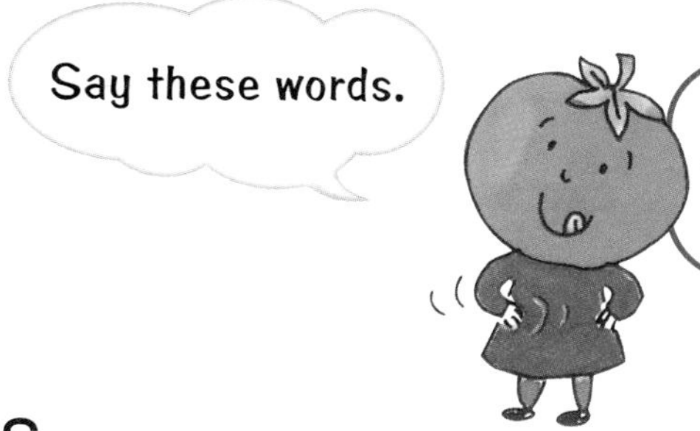

Say these words.

Circle the pictures that show equal parts.

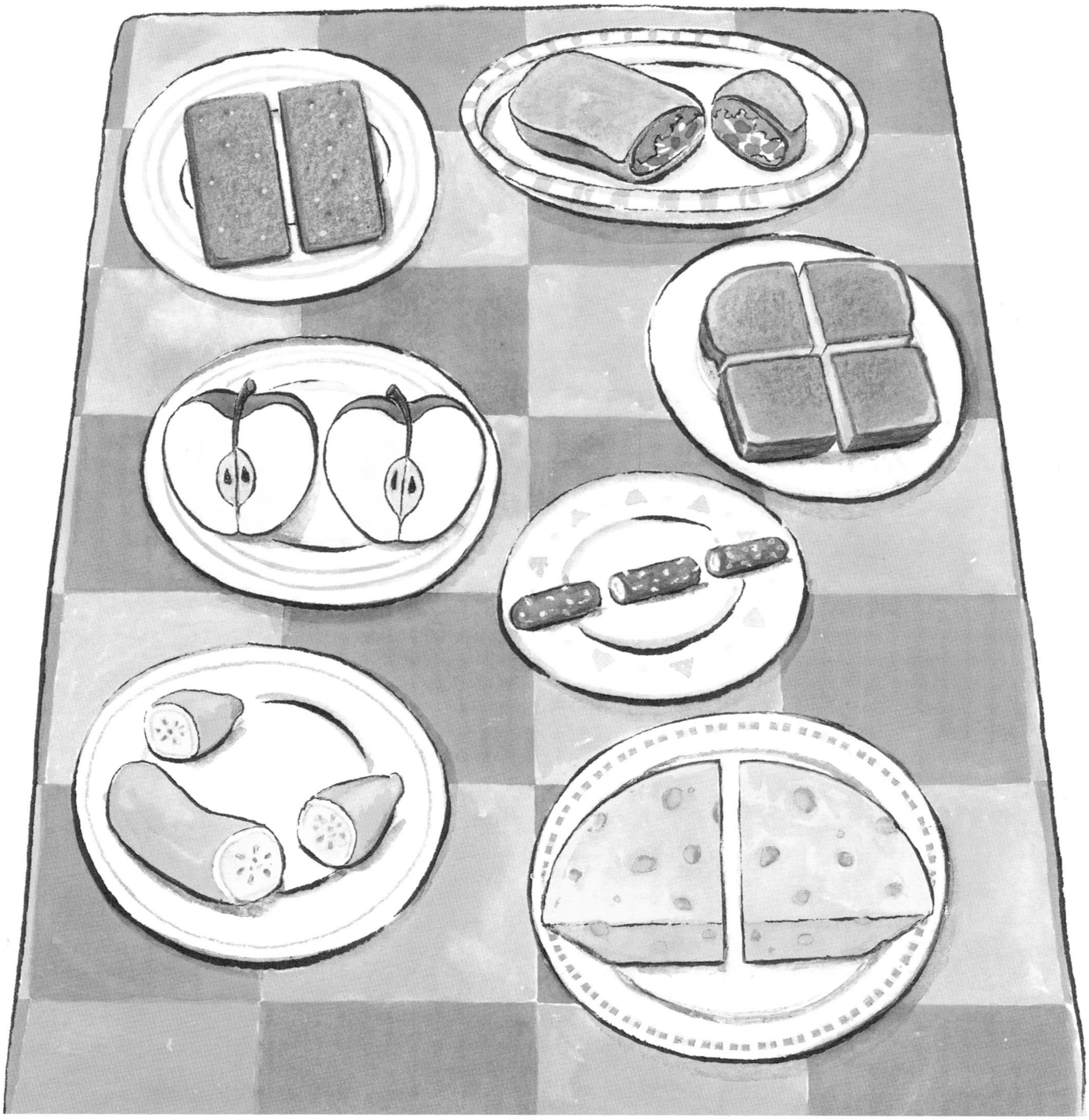

Math at Home: Your child is learning about equal parts.
Activity: When preparing a snack at home, ask your child to help you make equal parts of a food item.

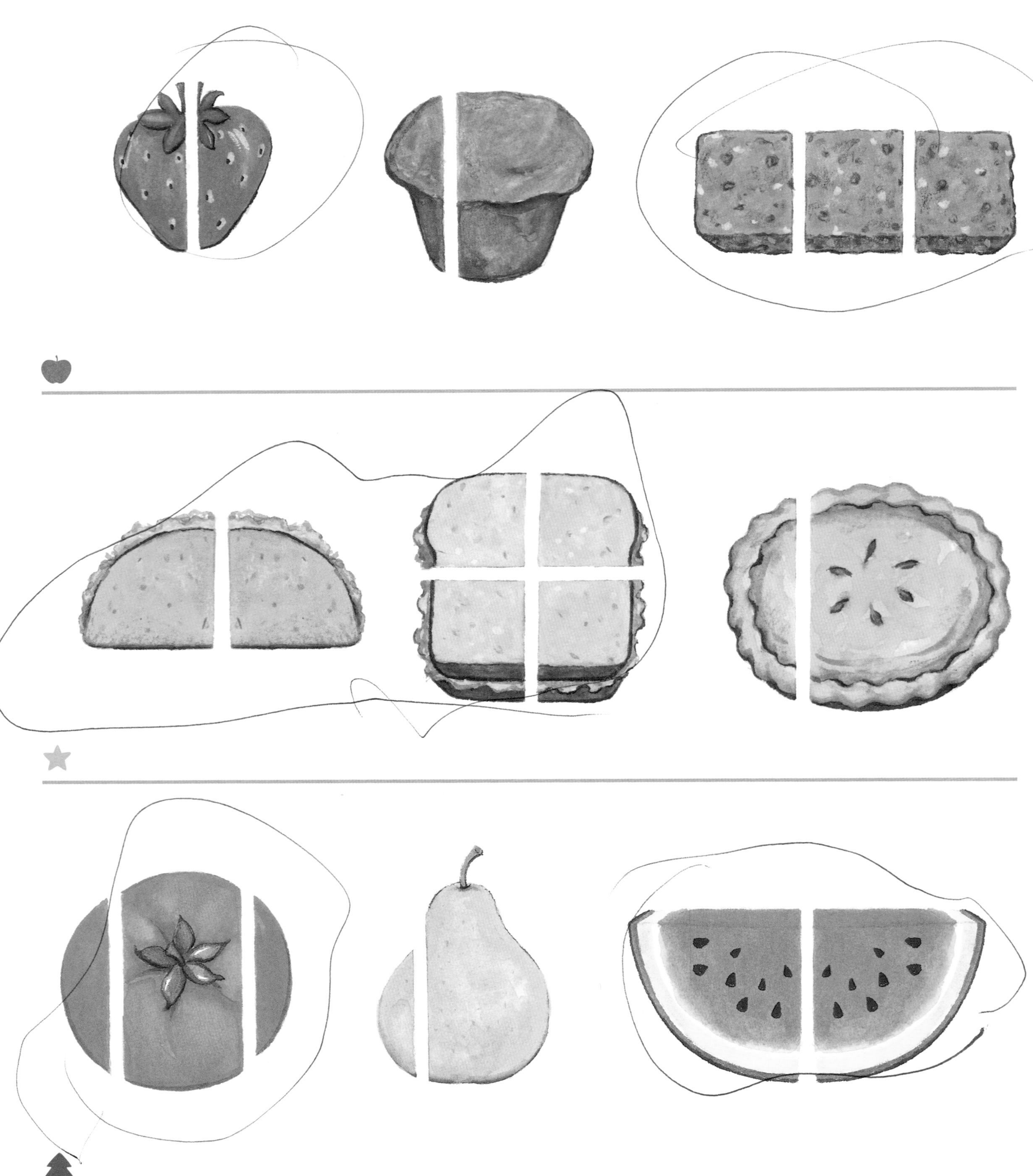

Circle the pictures that show equal parts.

Name ____________________

halves one half

Circle the foods that show halves.

Math at Home: Your child identified halves.
Activity: Ask your child to tell how he or she knows which pictures on the page show halves.

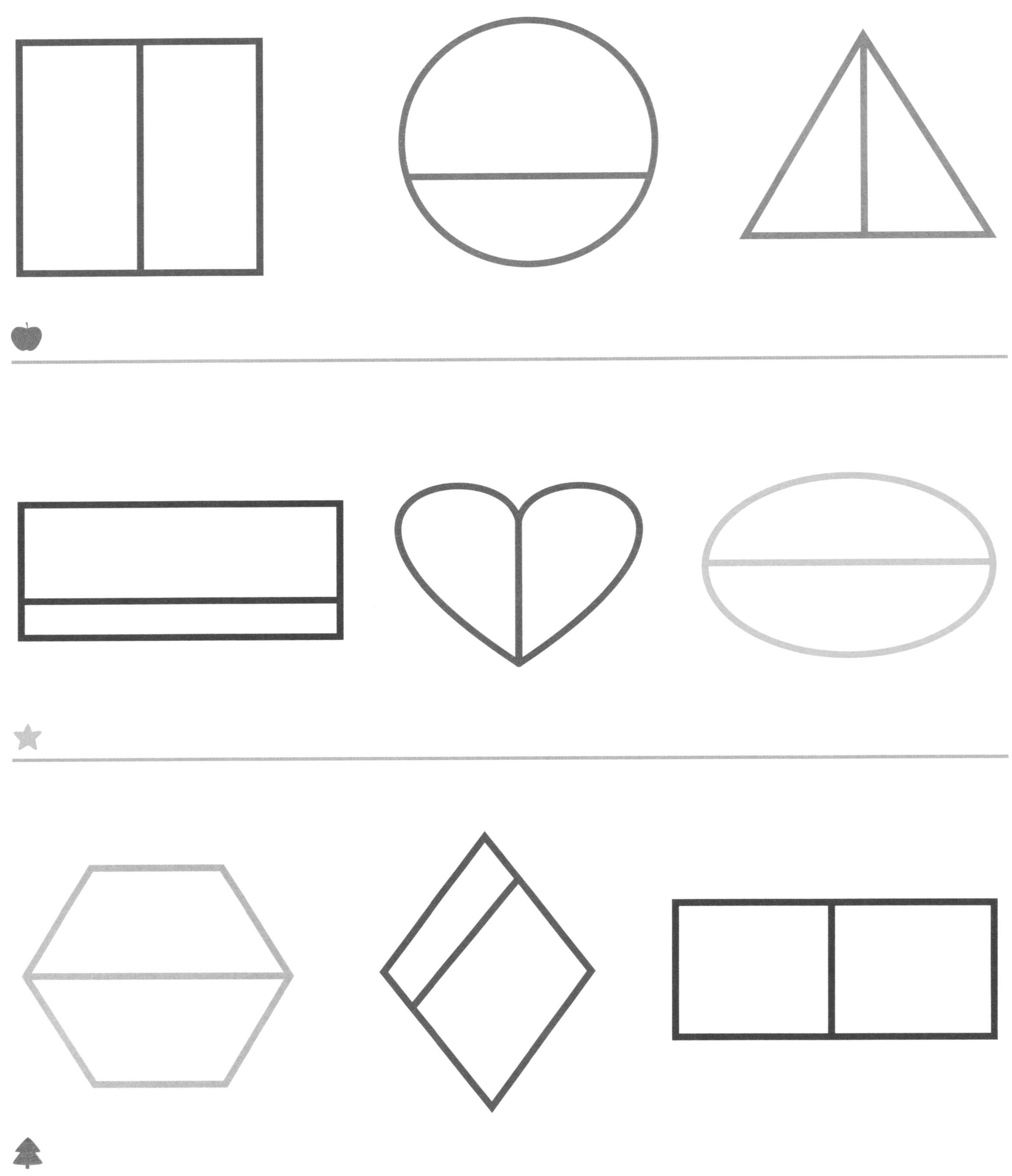

Find the shapes that show halves. Color one half red.

Name ____________________

Say these words.

Math Words
equal groups

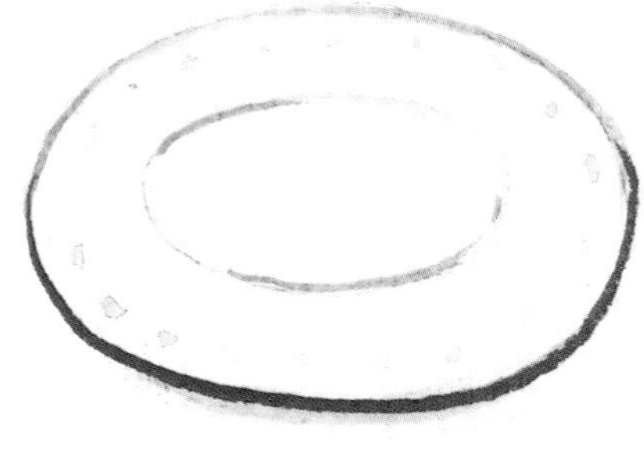

Use counters to show the food. Make equal groups. Draw pictures to show the equal groups.

Math at Home: Your child made equal groups.
Activity: At snack time, ask your child to put the snacks in equal groups.

Circle the pairs of plates that show equal groups.

Name ____________________

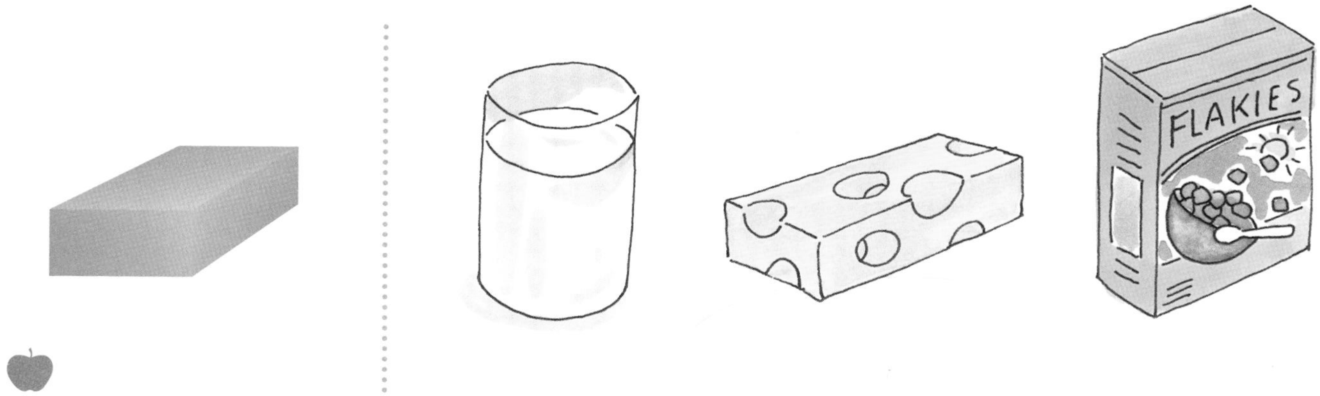

FLAKIES

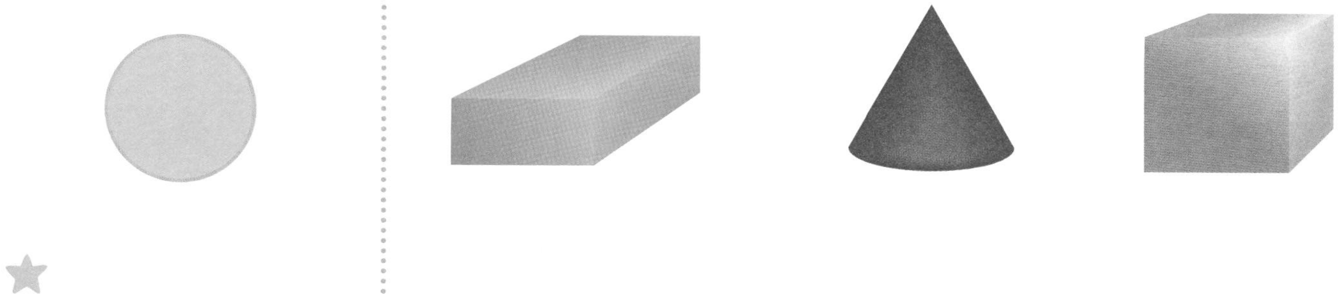

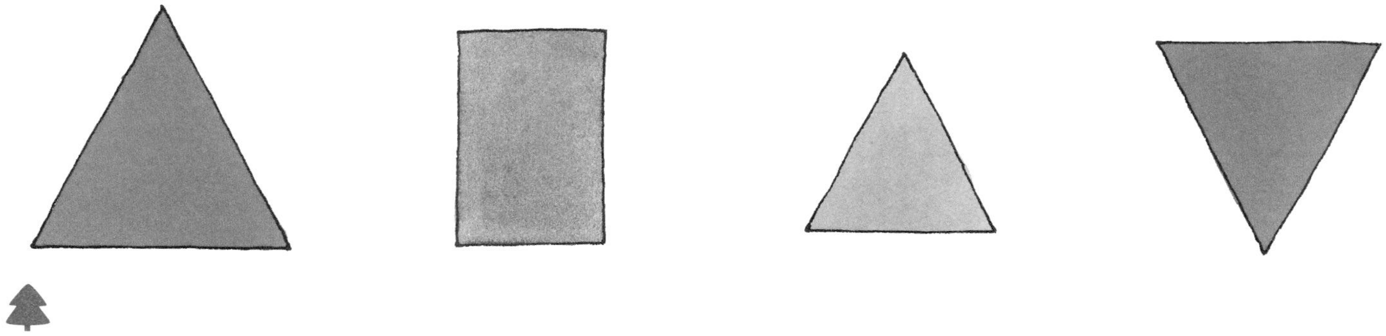

- Circle the items that have the same shape as the solid figure.
- Circle the solid figure you can use to make the plane figure.
- Circle the figures that have the same shape.
- Color to copy the pattern.

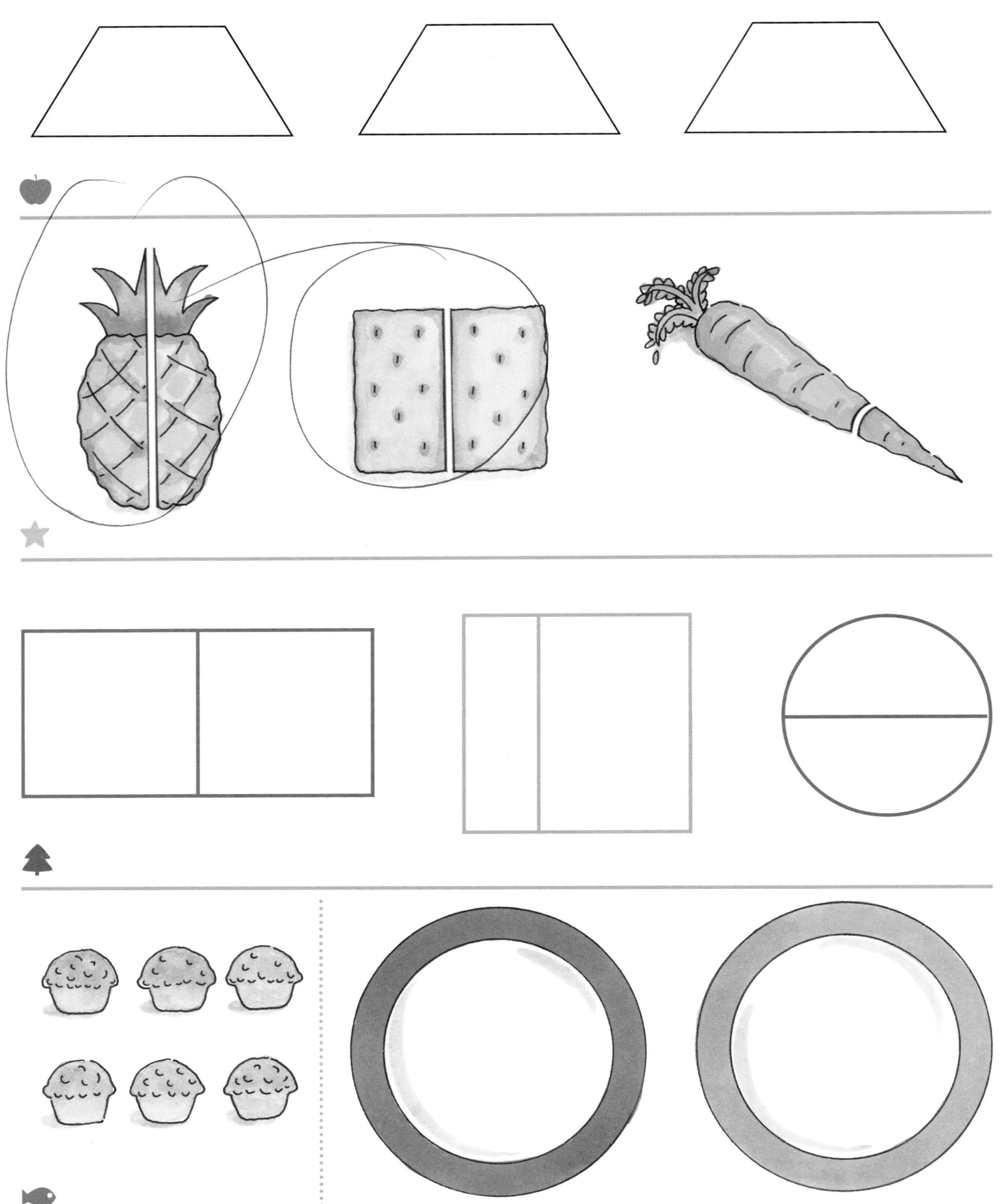

- Use pattern blocks to make the shape in three different ways. Color to show the pattern blocks.
- Circle the foods that show equal parts.
- Find the shapes that show halves. Color one half red.
- Draw pictures to show equal groups.

Picture Glossary

add

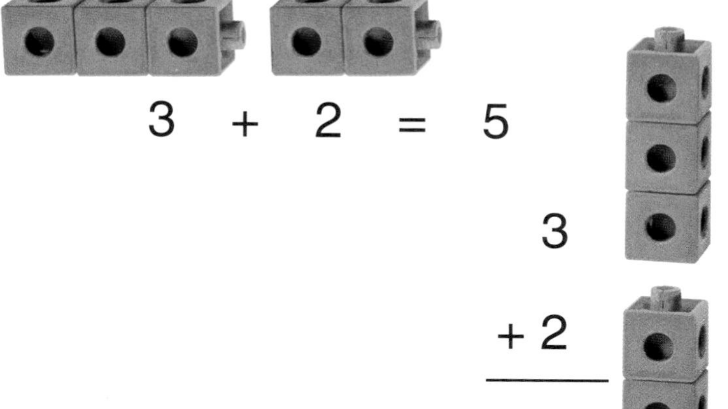

3 + 2 = 5

$$\begin{array}{r} 3 \\ +\,2 \\ \hline 5 \end{array}$$

always

You will always pick a .

after

6 7

↑ just **after** 6

bar graph

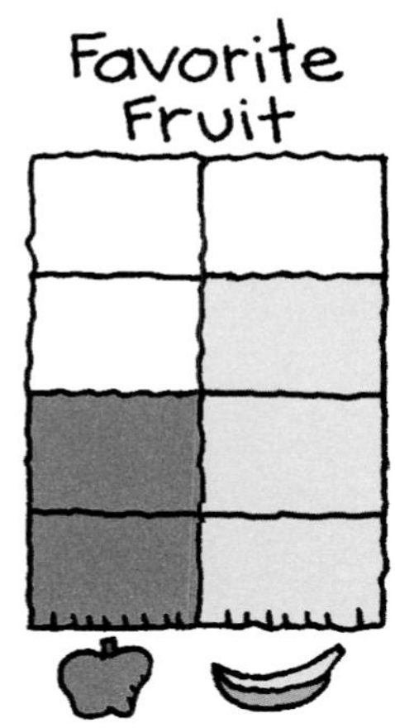

A graph using colored bars.

all

All of the cubes are in the cup.

before

4 5

↑ just **before** 5

Picture Glossary

behind

The boy is behind the desk.

calendar

calendar

between

5 6 7

↑ **between** 5 and 7

cent ¢

1¢ 1 cent

bottom

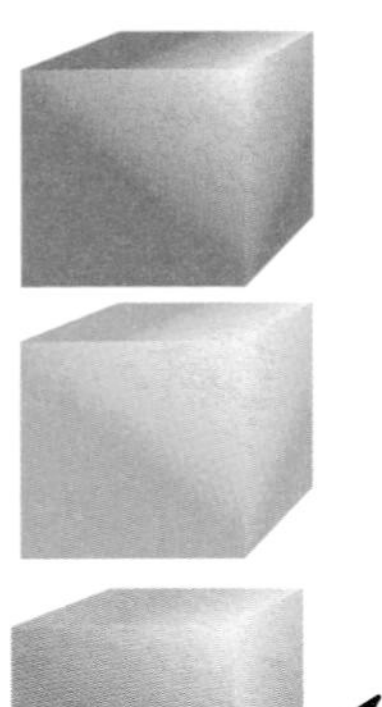

← bottom

circle

cold

cube

cone

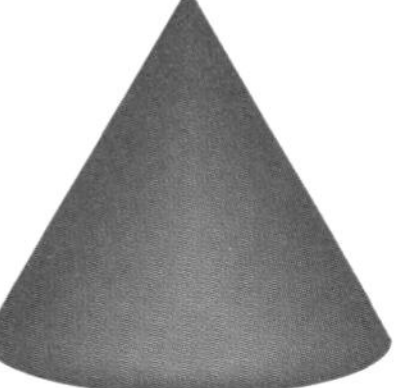

cylinder

count

1, 2, 3

Count to find how many.

day

September

day →

Sun	Mon	Tues	Wed	Thurs	Fri	Sat
					1	2
3	4	5	6	7	8	9
10	11	12	13	14	15	16
17	18	19	20	21	22	23
24	25	26	27	28	29	30

Picture Glossary

difference

$$\begin{array}{r} 5 \\ -2 \\ \hline 3 \end{array} \leftarrow \text{difference}$$

The **difference** tells how many are left.

dollar

$1.00

100¢

different

different

equals

$3 + 2 = 5$

equals

dime

10¢

10 cents

equal groups

equal groups

equal parts

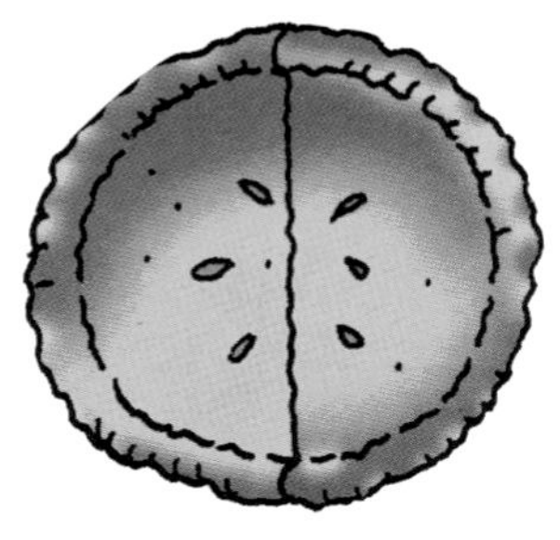

2 equal parts

first

first

far

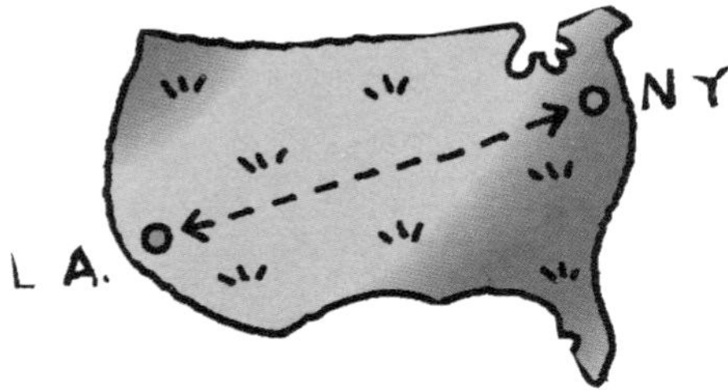

far

halves

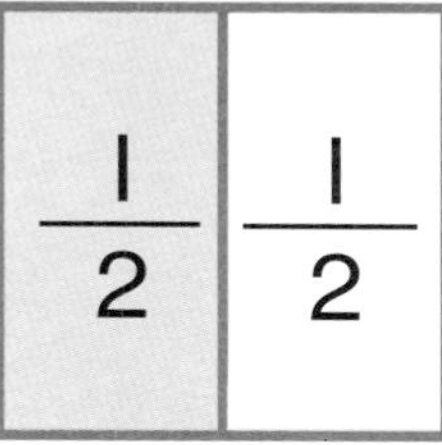

2 equal parts

fewer

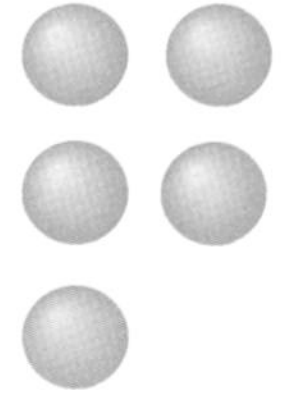

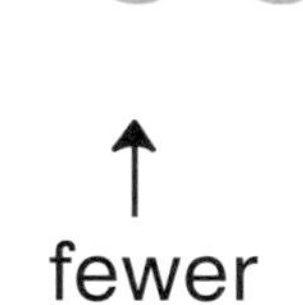

fewer

heavier

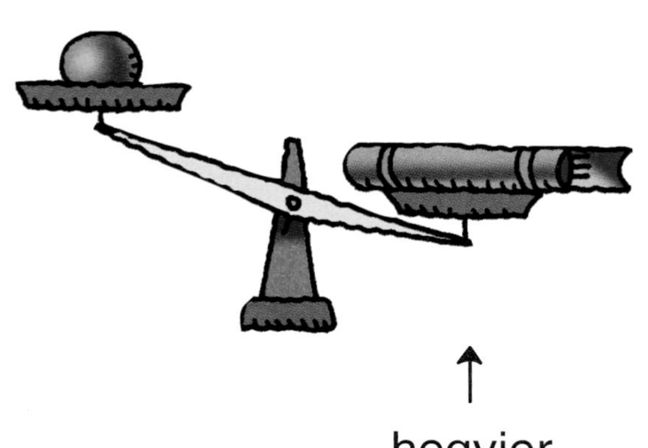

heavier

hot

hot

last

last

hour hand

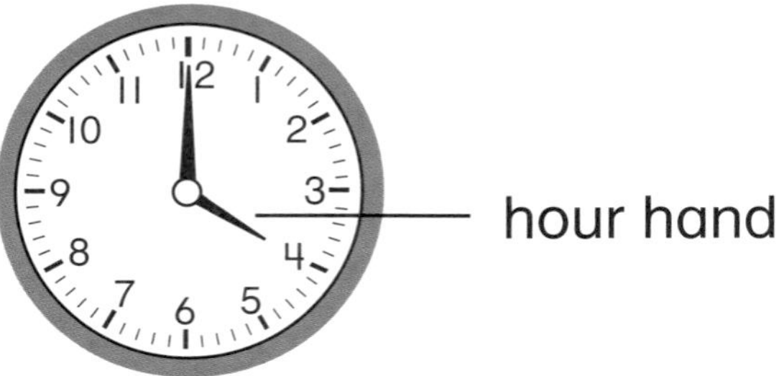

left

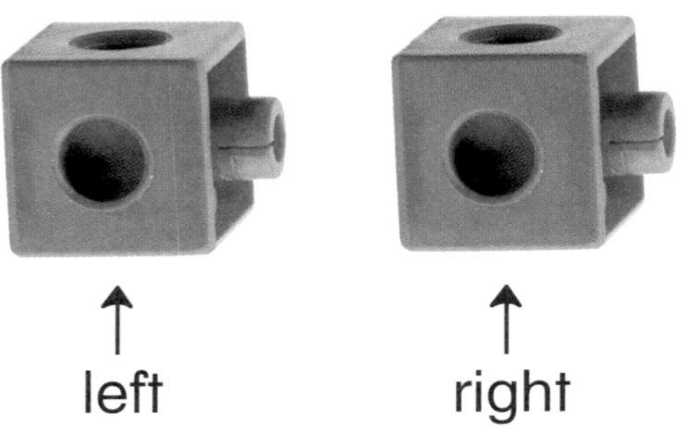

left

right

inside

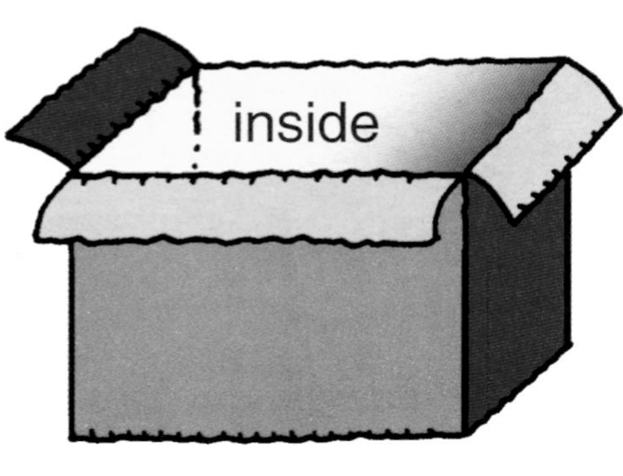

outside

lighter

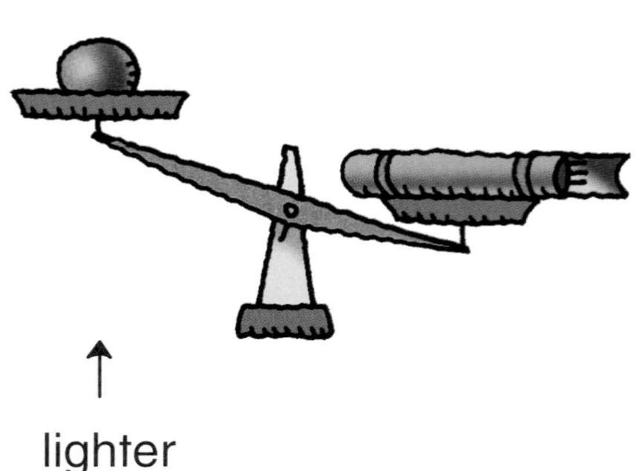

lighter

Picture Glossary

longer

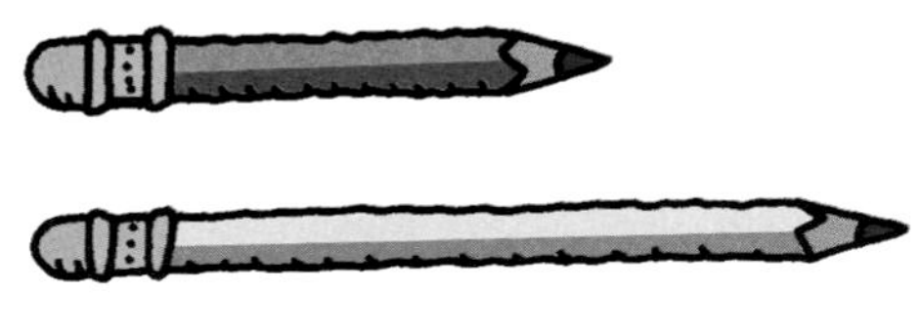

longer

minus

$5 - 2 = 3$

minus

maybe

Maybe you will pick a .

minute hand

middle

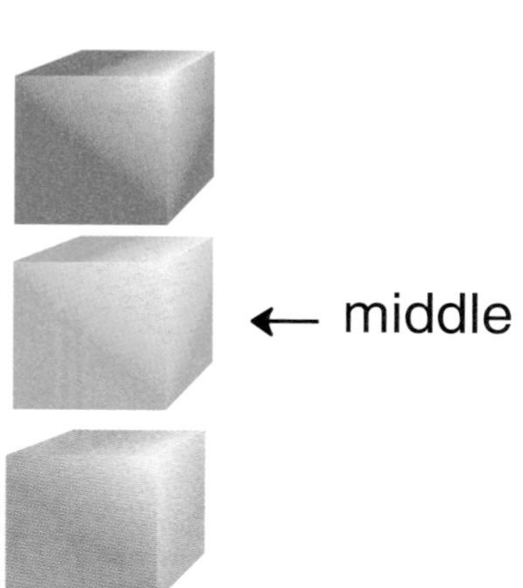

month

more

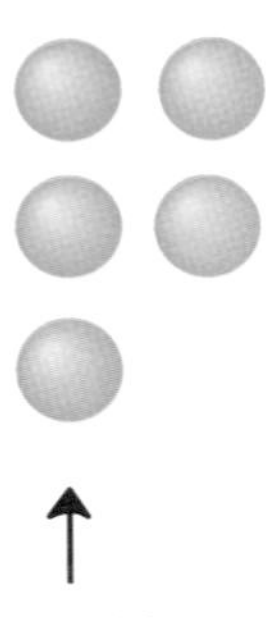

more

next

next

near

near

nickel

5¢ 5 cents

never

You will never pick a .

night

night

none

None of the 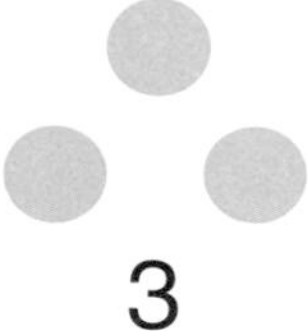 are in the cup.

ones

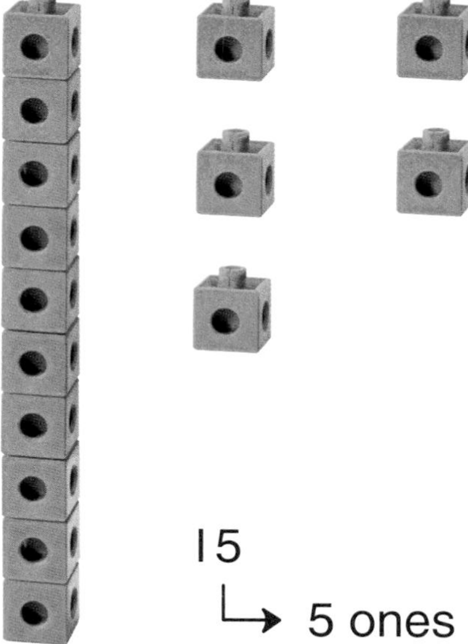

15
↳ 5 ones

number

3

A **number** tells how many.

order

0, 1, 2, 3, 4, 5

These numbers are in **order**.

one half $\frac{1}{2}$

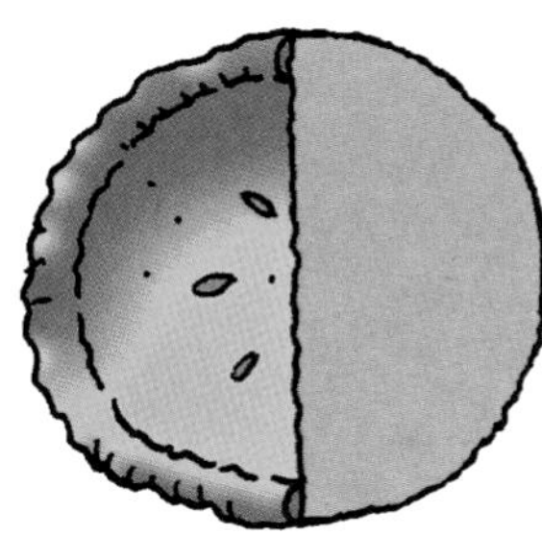

$\frac{1}{2}$

outside

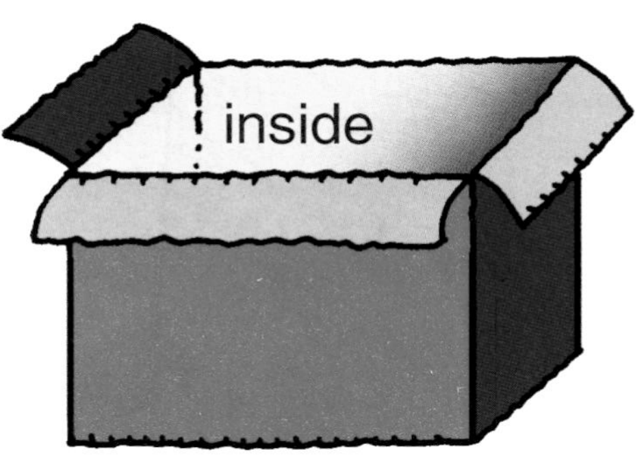

outside

over

The frog is over the table.

picture graph

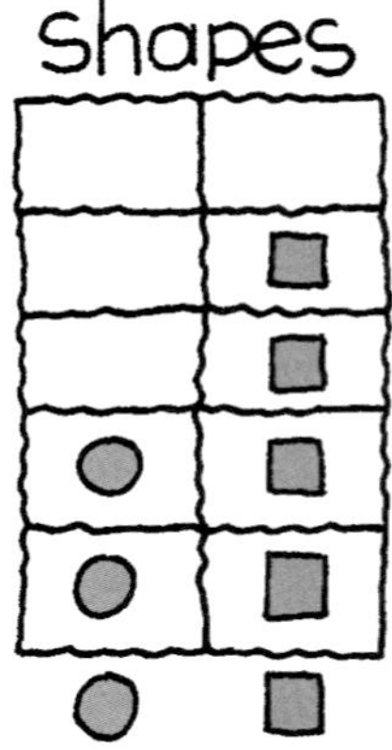

A **graph** using **pictures.**

pattern

plus

3 + 2 = 5

↑

plus

penny

1¢ 1 cent

quarter

25¢ 25 cents

Picture Glossary

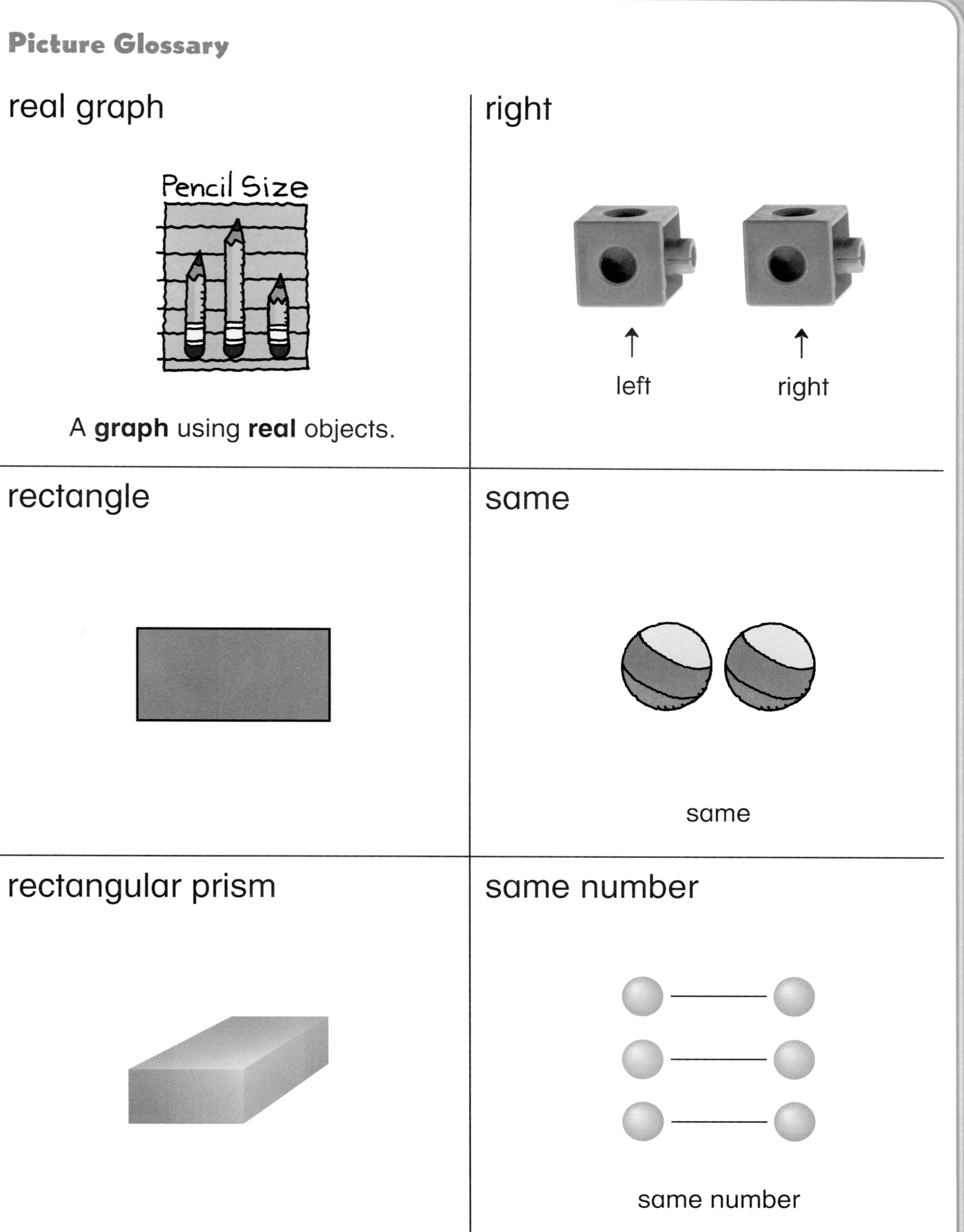

Picture Glossary

shorter

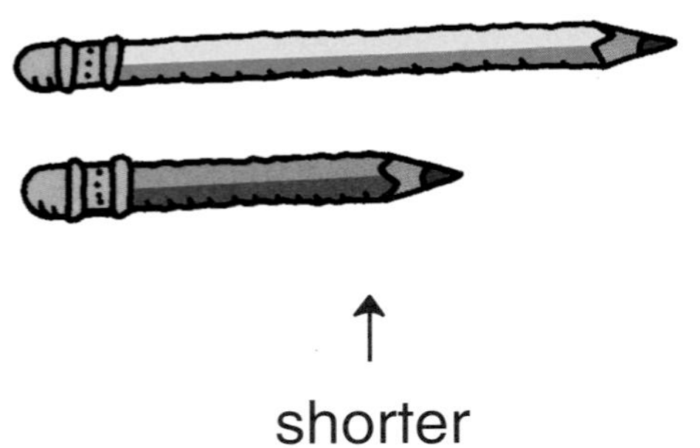

shorter

sort

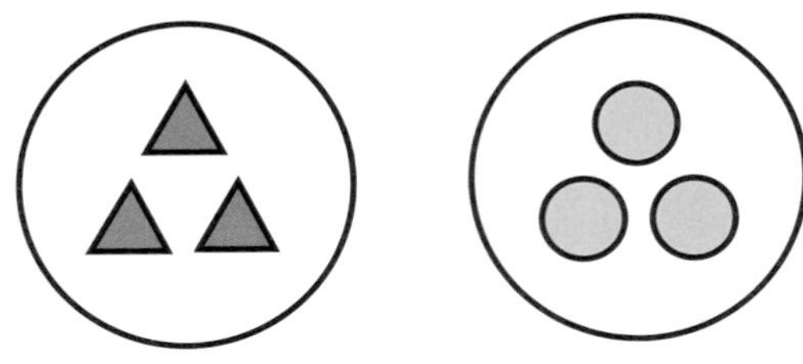

These items are **sorted** by shape.

skip count

2 4 6 8 10

sphere

some

Some of the cubes are in the cup.

square

Picture Glossary

subtract

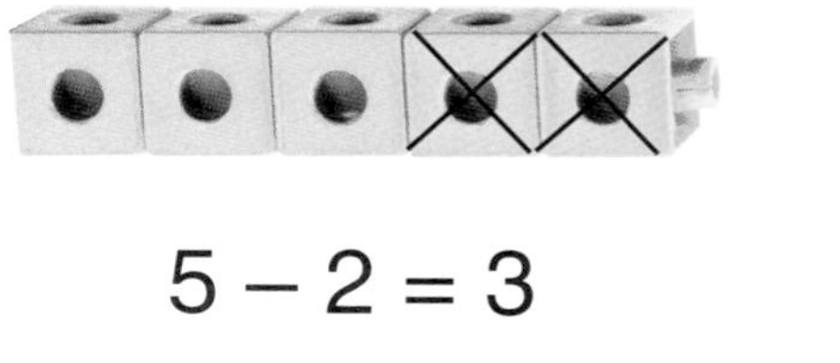

5 – 2 = 3

$$\begin{array}{r} 5 \\ -\,2 \\ \hline 3 \end{array}$$

sum

3 + 2 = 5 ← Sum

The **sum** tells how many there are in all.

table

tally marks

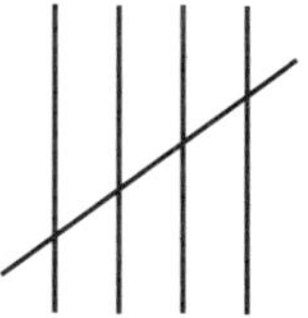

Tally marks are a way to keep track of counting.

tens

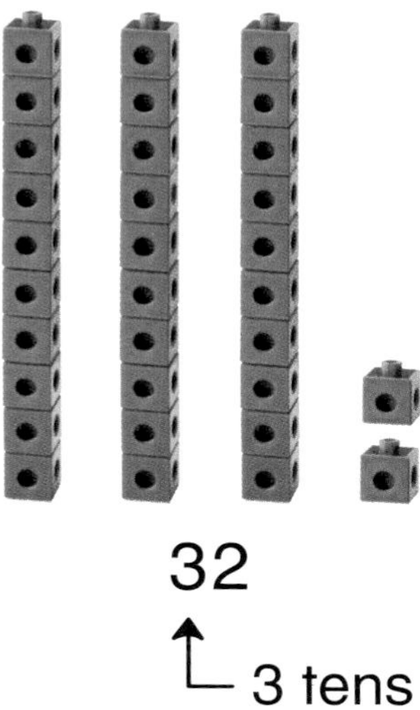

32

↑ 3 tens

today

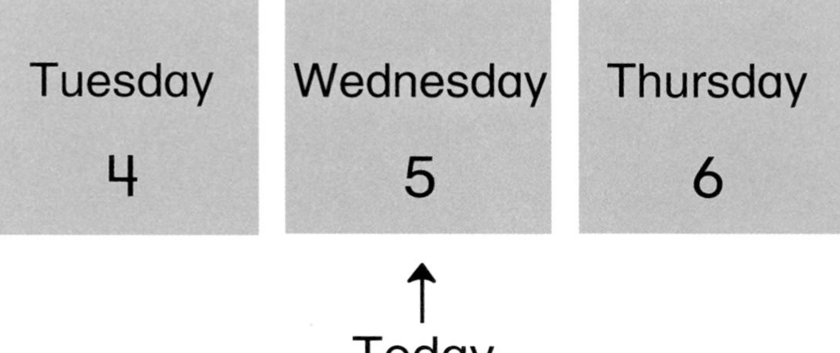

↑
Today

tomorrow

Tuesday	Wednesday	Thursday
4	5	6

↑ Tomorrow

under

The frog is under the table.

top

← top

yesterday

Tuesday	Wednesday	Thursday
4	5	6

↑ Yesterday

triangle

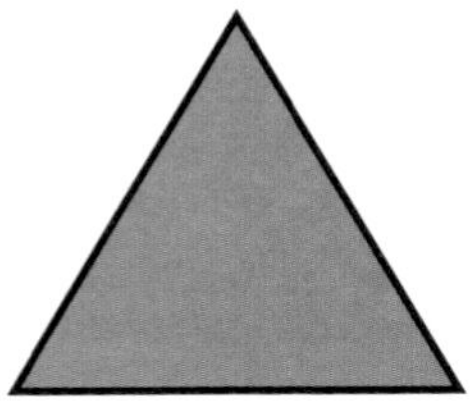

Credits

PHOTOGRAPHS

All photographs are by McGraw-Hill School Division (MHSD) and Mike Gaffney for MHSD; Michael Provost for MHSD; Dave Mager for MHSD.

ILLUSTRATIONS

Yvette Banek: 146, 148, 149, 150, 151, 153, 157, 158; Karen Bell: 221, 222, 223, 226, 227, 243, 244; Ka Botzis: 19, 21, 23, 26, 27, 29, 31; Cindy Brody: 34, 37, 38, 40, 41; Karen Stormer Brooks: 57, 59, 60, 61, 65, 69, 71, 75, 76; Annette Cable: 32, 33, 35, 36, 39, 42; Cardona Studio: 181, 182, 183, 184(top), 185, 186(top), 188(top); Chi Chung: 21, 22, 28; Diana Craft: 162, 167, 168, 170, 172; Lynne Cravath: 1, 3, 4, 5, 6, 7, 9, 11, 12, 13, 14, 15, 17; Jane Dippold: 245, 246, 247, 248, 251, 256, 257, 259, 261, 262, 264; Jim Durk: 177, 179, 183; Rusty Fletcher: R1, R3, R5, R6, R7, R8, R9, R10, R11, R13, R14; Susan Hall: 2, 3, 5, 8, 9, 10, 13, 16, 18; Holly Hannon: 100, 102, 104, 106, 108, 110, 111, 112, 115, 116, 119, 121, 122; Pat Hoggan: 78, 80, 82, 84, 85, 90, 96; Gay Holland: 20, 24, 25, 30, 32; Sharon Holm: 189, 191, 192, 197, 198, 199, 201, 215, 216; Anthony Lewis: 145, 147, 149, 151, 152, 154, 155, 156, 157; Ben Mahan: 162, 163, 164, 165, 166, 169, 171, 173, 174, 175, 176; Lynn Martin: 217, 219, 220, 224, 225, 228, 237, 238; Scott McDougal: 123, 125, 129, 130, 135, 139, 141, 142, 144; John Mengel: 192, 193, 194, 198(bottom), 200(bottom), 201, 202(bottom); Mas Miyamoto: 44, 50, 52, 55, 56; Susan Nethery: 43, 45, 46, 47, 51, 54; Laura Ovresat: 58, 62, 63, 64, 68, 69, 70, 71, 72, 73, 74; Susan Parnell: R2, R3, R4, R5, R9, R11, R12; Stephanie Peterson: 190, 193, 194, 195, 196, 200, 209, 210; Lou Pollack: 249, 257, 258, 259, 261, 262; Chris Powers: 88, 89, 92, 94, 97; Mike Rangner: 99, 103, 113, 114, 117, 118, 119, 120; Stephanie Roth: 77, 79, 81, 87, 91, 93, 95, 98; Bob Staake: 124, 126, 128, 133, 134, 137, 138, 143.